SOUNDWALKING

Soundwalking brings together a diverse group of contemporary scholars, artists and thinkers in one of the first comprehensive studies of soundwalking – the practice of moving through space while carefully listening to what it has to say – to address urgent challenges and concerns of an environmental, ethical, social and technological nature.

Besides gaining insight into the historical development of soundwalking as a scholarly method and artistic genre, the reader will have a chance to learn from emerging voices concerned with this practice, of many different backgrounds and positionalities. *Soundwalking* demonstrates how attentive listening and walking might help with more careful and responsible navigation through the complex dimensions of our shared environments and entangled histories, often imperceptible on a day-to-day basis. The book encourages scholars, artists, and also those unfamiliar with the concept, to engage with it in their respective fields and subjects of interest as an interdisciplinary method of critical inquiry and a creative mode of communication.

This book inspires readers to discover anew the potential of walking and listening, and will be of interest to students, researchers and practitioners in the areas of studies directly concerned with sound and beyond, including environmental humanities, arts, design, landscape architecture, media, and cultural studies.

Jacek Smolicki is an interdisciplinary artist, researcher, designer and educator. His work brings historical, critical and existential dimensions to listening, recording and archiving practices in human and more-than-human contexts. He holds a PhD in Media and Communications from Malmö University and is currently a postdoc researcher at Linköping University. He is also a guest researcher at Uppsala University, Simon Fraser University, and 2022/2023 Fulbright visiting scholar at Harvard. In 2019, he co-founded the Walking Festival of Sound.

SOUNDWALKING

Through Time, Space, and Technologies

Edited by Jacek Smolicki

Routledge
Taylor & Francis Group
LONDON AND NEW YORK

Cover image: © Jacek Smolicki Imprint

First published 2023
by Routledge
4 Park Square, Milton Park, Abingdon, Oxon OX14 4RN

and by Routledge
605 Third Avenue, New York, NY 10158

Routledge is an imprint of the Taylor & Francis Group, an informa business

British Library Cataloguing-in-Publication Data
A catalogue record for this book is available from the British Library

ISBN: 978-1-032-04424-8 (hbk)
ISBN: 978-1-032-04422-4 (pbk)
ISBN: 978-1-003-19313-5 (ebk)

DOI: 10.4324/9781003193135

Typeset in Bembo
by MPS Limited, Dehradun

Access the companion website: www.para-archives.net/soundwalking

CONTENTS

CONTRIBUTORS

Usue Ruiz Arana is a landscape architect, researcher and educator at Newcastle University. Usue's research is focused on two interrelated strands: more-than-human conceptualisations and listening as a methodology for Landscape Architecture practice. Through her research, Usue seeks to affirm non-humans as designers, and design and art as forms of research, drawing from her experimental arts methods that include soundwalks, photography and temporary installations, and a 20-year career in practice.

Elena Biserna is an independent researcher and occasional curator based in Marseille, France. She writes, talks, facilitates workshops or collective projects, curates and sometimes performs. Her interests are focused on listening and on contextual, time-based art practices in relationship with urban dynamics, socio-cultural processes, the public and political sphere. Her writings have appeared in several publications (Les Presses du Réel, Mimesis, Le Mot et le Reste, Errant Bodies, Amsterdam University Press, Cambridge Scholar, Castelvecchi, Bloomsbury, etc.) and journals. She has recently edited: *Walking from Scores* (2022) and *Going Out. Walking, Listening, Sound-Making* (2022).

Louis Chude-Sokei teaches at Boston University and directs the African American Studies Program. Books includes the award-winning *The Last Darky: Bert Williams, Black on Black Minstrelsy and the African Diaspora* (2005), *The Sound of Culture: Diaspora and Black Technopoetics* (2015) and the acclaimed memoir, *Floating in A Most Peculiar Way* (2021). He is Editor in Chief of *The Black Scholar*, and founder of the sonic art and archival project, Echolocution. Other projects include collaborations with electronic music icons, Mouse on Mars, and legendary choreographer Bill T. Jones. Chude-Sokei was also a curator of Carnegie Hall's 2022 Festival of Afrofuturism.

Milena Droumeva is an Associate Professor of Communication and Sound Studies at Simon Fraser University specialising in mobile technologies, sound studies and multimodal ethnography. They have a background in acoustic ecology and work across the fields of urban soundscape research, sonification for public engagement, as well as gender and sound in video games. Milena's current research combines livability with urban soundscapes research, including sonification of socially relevant data.

Patrick Farmer lives on the Malvern Hills and teaches in Oxford, where he is also the manager of the Sonic Art Research Unit at Oxford Brookes University and a curator of the audiograft festival. Farmer has published several books and written compositions for the Extradition Series and the Set Ensemble. A monaural artist, he is part of an AHRC funded project, Tinnitus, Auditory Knowledge and the Arts, and has recently had essays published online by Zeno Press, Riffs Journal and Socrates on the Beach.

Amanda Gutiérrez (b. 1978, Mexico City) explores the experience of political listening and gender studies by bringing into focus soundwalking practices. Trained and graduated initially as a stage designer from The National School of Theater, Gutiérrez uses a range of digital media tools to investigate everyday life aural agencies and collective identities. Approaching these questions from aural perspectives continues to be of special interest to Gutiérrez, who completed her MFA in Media and Performance Studies at the School of the Art Institute of Chicago. She is currently elaborating on the academic dimension of her work as a PhD student in Arts and Humanities at Concordia University.

Katt Hernandez is a violinist, composer and improvisor. She began her PhD, *The Ephemeral City: Songs for the Ghost Quarters*, at Lund University in 2015. She has also worked with Klas Nevrin's Music in Disorder, Fire! Orchestra, GGR Betong, Lotte Anker's sub-habitat, among others. Her work has been featured recently at Antenn Festivalen, Datscha Radio, BlowOut, PULS festival of multichannel music, Nordic New Music Days, High Zero and more. Based in Sweden, Katt is also a veteran of experimental music scenes on the East Coast, alongside a vast array of musicians, dancers, visual artists, puppeteers, film makers and performance artists.

Paola Cossermelli Messina (BA, Sarah Lawrence College/MA, The New School) is a PhD candidate in Ethnomusicology at Columbia University, interested in soundings between Brazil and Lebanon. Past works include an ethnography on a jukebox in a lesbian bar and oral histories of Iranian women-identifying musicians in exile. She worked as a radio producer in São Paulo, where she was born, and is a sound designer by trade. Her work was presented at Harvard and Yale, and awarded by the Middle East Studies Association. Columbia's Center for Palestine Studies selected her as a Public Humanities Radio Play Fellow (2021–22).

Gascia Ouzounian is Associate Professor of Music at the University of Oxford. She is the author of *Stereophonica: Sound and Space in Science, Technology, and the Arts* (2021) and numerous articles on experimental music and sound art, sound technologies and sonic urbanism. Recent projects include *Scoring the City* (2020–ongoing), which explores experimental notations for urban design, and *Acoustic Cities: London and Beirut* (2019), for which ten artists created works responding to the acoustic conditions of London and Beirut. At Oxford, Ouzounian leads the European Research Council-funded project Sonorous Cities: Toward a Sonic Urbanism (soncities.org).

Tim Shaw's practice is concerned with the many ways people listen, specifically how listening environments can be constructed or explored using a diverse range of techniques and technologies. He has a background in recording sound, and his practice is anchored in the creative use of field recordings. He works with electronics, video, synthesis, sound objects, self-made hardware and DIY software. Tim presents work in galleries, festivals, museums, through residencies and cultural events nationally and internationally. Tim is a Lecturer in Digital Media at Newcastle University where he also holds a PhD.

Jacek Smolicki, PhD, is an interdisciplinary artist, designer and researcher. His work explores historical, critical and existential dimensions of listening, recording, and archiving practices in human and more-than-human realms. His work takes the form of soundwalks, soundscape compositions, diverse forms of writing, experimental archives, and audio-visual installations. He has performed, published and exhibited internationally, and in 2019 co-founded the Walking Festival of Sound. He is currently an international postdoc at Linköping University with guest affiliations at Simon Fraser University in Vancouver, Harvard University and Uppsala University.

ACKNOWLEDGEMENTS

This book would not have come about without an inspiring ecosystem of artists, thinkers and researchers interested in critical, reflexive and inventive potential of walking and listening. I would particularly like to thank Barry Truax for his kind invitation to visit the Sonic Research Studio at Simon Fraser University in Vancouver, BC as a guest researcher in 2020 and 2022. Thank you Barry for a series of inspiring conversations and guidance through the history and presence of soundscape studies as pursued in and beyond Canada. Initiating and editing this book has been one of many components to my international postdoctoral research into soundwalking, field recording and listening practices in the context of arts, environmental humanities and philosophy of technology. This independently initiated project was funded by the Swedish Research Council and, thanks to support from Jesper Olsson, between 2020 and 2023 was situated at the Culture and Society Department of Linköping University in Sweden. Thank you Jesper for your enthusiasm, and for consistently making sure that the doors of academia are ever-more widely open to artistic thinking and making. My sincere words of gratitude go to Hildegard Westerkamp, the key figure to soundwalking, who needs no introduction. Thank you for invaluable conversations that, despite the constraining forces of the Covid-19 pandemic, we had a chance to weave during my visits to Vancouver. My warm thanks to Giorgio Magnanensi, the artistic director, and Heather McDermid, the curator at Vancouver New Music, who have consistently been making room for soundwalking artists as important contributors to the local culture. Thank you Giorgio, for your support and kind introduction to your network. Big thanks to Tim Shaw for those stimulating years of collaboration. Together with Tim, in 2019 I co-founded and ever since co-directed the Walking Festival of Sound. It is through the work towards several editions of this nomadic initiative that I got to know and meet some of the scholars and artists who then became contributors to this book. Many thanks to

Yvette Janine Jackson for her kind invitation to join the environment of Department of Music at Harvard as a guest artist and researcher during the 2022/2023 term. Words of gratitude to Milena Droumeva from Simon Fraser University for inviting me to be part of the regularly assembling sonic crew that constructively builds on the legacy of the Sonic Research Studio. Words of appreciation to Milena, Meri Kytö and Sławomir Wieczorek for the commentary on respective parts of this book. Many thanks to Hannah Rowe, editor at Routledge, and Emily Tagg, editorial assistant, for their professional guidance on every phase of this project. Words of appreciation to Sarah Webb for her detailed and attentive copyediting work, and Jennifer Hicks for her supervision of the production phase. Last but not least, I am grateful to all contributing authors. Thank you for your patience, dialogues and exchanges throughout this journey, but, foremost, for believing in this project and being willing to share your diligent work and passion with the readers of this book.

INTRODUCTION

Jacek Smolicki

It is Sunday, July 17, 2022, 12:30 pm. With my 3-year-old daughter, I am standing at a little square near Roundhouse community Arts Center in Vancouver, waiting to take part in a soundwalk, a local event celebrating so-called World Listening Day.[1] For about an hour, we will walk and listen to the neighborhood while remaining silent. Temperature is moderate. Clouds begin to cover the sky. It is not raining, although it is rather obvious it soon will. The square is regularly crossed by people, mostly young families heading to the community center or the playground nearby. The soundscape is permeated by noises from the nearby car traffic, occasional ambulance signals, kids laughing, the clattering of strollers and bicycles against the square's cobblestones. All these sonic events are spanned by a sustained web of human voices radiating primarily from participants who gradually gather to join the soundwalk. Initially, there seems to be nothing special about the way the place sounds. But the longer I am standing there and waiting, the more I am aware of the soundscape, or my experience of it.

Two years ago, during my previous trip to Vancouver as a guest postdoc at Simon Fraser University, the stronghold of soundscape studies, social gatherings of this kind were radically constrained by the expanding Covid-19 pandemic. Schools, and cultural and community centres shut down. Public gatherings and events were cancelled, even banned. It is not that by now the pandemic has ended altogether. In fact, at the moment of writing this introduction, new Covid variants and prospects of other waves rise on the horizon. Nevertheless, the situation now, at least here in Vancouver, seems much different from only two years ago. And this gets expressed quite literally in how the city sounds as well as in how one listens to it – its at once old and renewed kind of aurality.[2] While standing at the square and waiting to take part in a collective soundwalk, I feel like I am already soundwalking: through my memories, immediate and distant associations as well as visions of the near future.

DOI: 10.4324/9781003193135-1

The pandemic taught us, or reminded us, that nothing can be taken for granted. Despite technological advancements, predictive algorithms, and managerial power over almost any aspect of our planetary life, today, perhaps more than ever before, nothing can be granted the status of certainty. In the course of several days or so, the most deeply rooted, familiar and normalized everyday life activities might radically transform or disappear altogether. And vice versa, new, unexpected routines and practices quickly take over those habituated ones, fill in the void, making many of us feel uncomfortable, at least at the beginning, and yet hopeful that what we presently experience is only a temporary diversion from the preceding state of affairs. What many of us considered (or were privileged to treat as) a state of exception, becomes a permanent norm. As many have rightly pointed out, the Covid pandemic – in its strangely hybrid state of being at once an exception and persistently settling norm – foregrounded a number of troubling issues that we, individuals and societies, had kept at the peripheries of our perception. And hearing. Either as a deliberate act or involuntarily, through structural decisions of authorities and power holders. Casting a new light on old problems – social and environmental injustice, racism, sexism, or ageism – this unanticipated situation demanded from us to rethink how we live. From us, those concerned with sound, it called to question how we listen. In this need for rethinking our ways of being and navigating the world, many of us saw the only glimmer of hope that the pandemic could possibly generate. In our surrounding soundscapes, or renewed awareness of their long ignored natural rhythms and sonorities, many of us wanted to hear the beginnings of a more balanced and just reality. However, this period, as we are witnessing in many places around the world, has taken people into numerous, often dark directions, many of which reinscribe rather than challenge those revealed inequalities. Instead of gaining traction, the new, or renewed opening of how we tend to each other and our planet seems to have been put on hold. But if possibilities of direct action might be impeded, there is still room for telling stories and generating insightful accounts. After all, as Joanna Zylinska rightly reminds us in her *Minimal Ethics for Anthropocene*, "stories have a performative nature: they can enact and not just describe things – even if there are of course limits to what they are capable of enacting" (Zylinska, 2014).

This book began from a set of questions in some ways resonant with the concerns mentioned above, although several months before the pandemic brought them into an even sharper relief. Can soundwalking, a seemingly trivial combination of walking and listening, be approached as a spatio-temporal condition for reconfiguring how we sense and act in the world today? Can attentive walking and listening unite into a technique for disrupting the familiar and familiarizing with the unknown, strange, uncanny, even inconvenient and monstrous?

The soundwalk I am about to take is offered by Soundwalk Collective. It is an initiative that emerged through a collaboration of people interested in exploring educational and social potential of listening to Vancouver's soundscapes. The collective has been a continuation of Hildegard Westerkamp's legacy generated through her work for the Vancouver Co-operative Radio in the late 1970s.

As many readers of this book are surely aware, Hildegard Westerkamp, a composer, writer, and key member of the World Soundscape Project (WSP) established in the 1960s by R. Murray Schafer at Simon Fraser University (Truax, 1974, 1999), has been a pivotal figure for the practice of soundwalking (McCartney, 2000, 2014; Sterne, 2019; Drever, 2020). Her work continuously inspires new generations of soundwalkers, including many contributors to this book. In those early years of her ongoing career, Westerkamp used radio as a channel for bringing the work pursued within the WSP context much closer to local communities and to open the potential of the practice up to broader audiences with no background in academia, sound studies, and composition. Through her weekly radio program entitled *Soundwalking*, creatively combining field recordings with spoken narrative, Westerkamp consistently encouraged people to listen to their environments. "Imagine radio," she wrote,

> that instead of numbing us to sounds, strengthens our imagination and creativity; instead of manipulating us into faster work and more purchasing, it inspires us to invent; instead of overloading us with irrelevant information and fatiguing us, it refreshes our acoustic sensitivity; instead of moving us to ignore thoughts and surroundings, it stimulates listening; instead of broadcasting the same things over and over again, it does not repeat; instead of silencing us, it encourages us to sing or to speak; to make radio ourselves; instead of merely broadcasting at us, we listen through it.
>
> *(Westerkamp, 1994, p. 89)*

Today, soundwalking is a widespread concept and practice, settling across many contexts and communities via multiple other formats and media. As John Drever writes in his recent study of diverse methodological applications of soundwalking, the richness of contexts in which it is being adopted proves that it is increasingly "emergent rather than a transplantable fixed practice" (Drever, 2020, p. 599). Magnifying this emergent and cross-disciplinary character of soundwalking is one of the aims of this book.

By 1978, when Westerkamp started her program, the terms soundwalk and soundwalking existed merely in relation to a method used by the WSP group to facilitate workshops on noise pollution. As Westerkamp explained in a conversation with me in Vancouver in the spring of 2020, her role at the time was to facilitate the group's soundwalks by preparing maps for analyzing noise pollution in specific locations. She would draw routes and designate stops at which workshop participants would be asked to respond to a questionnaire concerning levels of noise they witnessed. At this point, soundwalking functioned as a prescriptive method for generating research data, and thus was far from the ambitions that soon were to steer its further evolution, mostly due to Westerkamp's activist approach to sound, listening, and environment. Even though it continued only for one year, the Soundwalking program challenged conventional formats, subjects, and rhythms of radio production while also bringing the very medium closer to the realm of socially and environmentally concerned arts. Today, this interdisciplinary, boundary-breaking approach might not surprise anyone. Several decades ago, it surely did. And this quality did not apply to soundwalking alone, but to the whole field of acoustic ecology emerging at the time.

As Barry Truax, composer, sound scholar, and another key figure for acoustic ecology and WSP noted: at the time, the way sound and listening were approached as a subject of study was significantly constrained by perspectives and methods from acoustic sciences and engineering (Truax, 2019). Besides some exceptions, there was a troubling deficit of studies proposing more embodied perspectives on sound and listening, as culturally, communally, individually, or politically negotiated practices (ibid., p. 30). The interdisciplinary orientation that guided acoustic ecology attempted to address this imbalance. With the privilege of hindsight, one could say that the goal was achieved only partly. On the way to its realization, as the WSP members admit themselves, the project did not remain free from shortcomings and problems. One of the most frequently repeated critiques today points at problematic omissions of First Nations communities and ethnic minorities in the way the project attempted to portray key soundscapes and soundmarks of Canada. As Mitchell Akiyama argued: "the soundmarks they [WSP] so prized were deeply entangled with a silencing of Canada's indigenous population; of a protracted, often violent and brutal, campaign of assimilation that replaced one set of sonic practices with another" (Akiyama, 2015, n.p.)

Other undertakings, such as the unquestionably ambitious attempt to create consistent vocabulary for the realm of sound so it could enter academic discourse, in some cases resulted in dichotomies that keep haunting the field to this very day; Hi-Fi versus Lo-Fi sounds, for example (Schafer, 1977, p. 43; McCartney, 2016). From the perspective of contemporary debates, attempting to question human-exceptionalism, the premise of WSP might also seem problematic. While deeply engaged in environmental issues, the WSP scholarship and its methods were characterized by a strong human-centric orientation. The primary concern behind documenting, augmenting, soundwalking through, and organizing sound(scape)s was the well-being of humans, in particular those inhabiting urban areas. Critically evaluating the legacy of R. Murray Schafer from a local indigenous perspective and pointing to the influence of his soundscape compositions on environmentally oriented sonic arts at large, Dylan Robinson writes that "despite the growing eco-musicological discourse and influence of new materialism and non-representational theory on artistic practice, composition is often still created for the benefit of the human subject rather than for nonhuman others, including the land" (Robinson, 2020, p. 189). What all these different strands of critical reflection point to is above all the lack of recognition and critical assessment of specific situatedness of the project and positions from which it attempted to describe and assess aural environments. However, for today's generation of sound researchers and practitioners, these omissions, absences, and silences are increasingly clearer. They constitute a terrain for active engagement that in turn allows us to bring new life and significance to this increasingly relevant realm of scholarly and artistic undertakings (Droumeva and Jordan, 2019).

Unquestionably, sound and soundscape studies are in motion. And so are the ways we listen. Within the last decade alone we have been witnesses to a cascade of works following what some scholars denoted as a *sonic turn* in art and sciences (Drobnick, 2004; Augoyard and Torgue 2006; Atkinson, 2007; Kelly, 2011; Sterne, 2012). Moreover, a steadily growing number of sound studies across the world ever more boldly engage in generative and critical dialogs with other emerging perspectives and scholarly debates, new materialism, gender studies, environmental sciences and humanities, post-humanities, and post-colonialism among others (Järviluoma et al., 2009; Ochoa Gautier, 2014; Kanngieser, 2015; Droumeva and Jordan, 2019; Eidsheim, 2019; Steingo and Sykes, 2019; Kytö, 2019a; Kanngieser and Todd, 2020; Loveless, 2020; Ferrarini and Scaldaferri, 2020; Robinson, 2020; Lane and Carlyle, 2021; Losiak and Wieczorek, 2021; Wright, 2022). Resonating with this task of reevaluating the imperatives that have guided sound studies, including soundscape and acoustic ecology, the ambition of this book is to reassess motivations, aims, ontologies, and tools of soundwalking, one of the key practices through which those fields have been actualized, charted, sustained, and substantiated with knowledge. But before moving on to a concise walk-through of chapters in which contributors to this book offer particular enactments and alterations of the practice in question, let us take some steps back.

In an oft-quoted definition of soundwalking, Hildegard Westerkamp describes it as "any excursion whose main purpose is listening to the environment. It is exposing our ears to every sound around us no matter where we are" (Westerkamp, 1974, n.p.). As mentioned already earlier, in the last several decades the practice became subject to numerous theoretical, conceptual, and practical explorations. In many of them, soundwalking is set in dialog with other historical practices concerned with walking as a philosophical, aesthetic, and critical method for evaluating space (McCartney, 2014; Radicchi, 2017). Andra McCartney, sound artist and scholar who devoted much of her career to soundwalking, suggested that the practice should be

> situated in relation to long-standing artistic, philosophical and political concepts that theorize through the practice of walking, such as haiku poets' use of daily walks as a creative structure, writing about the figure of the flaneur and the situationist concept of the dérive, as well as the approaches of conceptual artists, such as those in the Fluxus movement.
>
> *(McCartney, 2014, p. 213)*

Broadening the definition and context, McCartney's work helped open the practice onto other fields. Consequently, in the last several decades, the practice's kinship with embodiment, mobility, place, subjectivity, and general alignment with qualitative and sensory approaches to understanding the world made it particularly prone to alliances with fields and debates already inclusive of those modes and concepts, as in humanities and arts (Smolicki, 2021), but also ones that sought to reconfigure their underlying imperatives and methodologies such as architecture, ethnography, anthropology, and urban planning (Kytö, 2019; Ruiz Arana, 2019).

With the emergence of technologies – increasingly portable and miniaturized sound equipment, recorders, microphones, and mobile phones – the definition of soundwalking expanded even more. Today, the term soundwalk is ever-more often used interchangeably with audio walk, listening walk, or audio guide.[3] This abundance of sound technologies and their availability for common audiences had an unquestionable impact on the way surrounding soundscapes are experienced by their listeners. For Truax, the value of soundwalks, traditionally unmediated in terms of technologies involved, has been to provide conditions for opening one's ears to "the most direct aural involvement possible with a soundscape" (Truax, 2012, p. 196). The question arises as to what incorporation of technological devices into soundwalking, such as microphones, headphones, hearing aids, and cochlear implants do to the listener? To what extent do they disrupt that "most direct aural involvement" that WSP scholars saw as the fundamental aspect of soundwalking? And, conversely, to what extent can technologies facilitate this access?

There are no certain answers to those questions because much depends on what ontological perspective we apply to define, understand, and evaluate the role of technologies in our diverse ways of being in the world. Following Bernard Stiegler, biological and technological evolutions are not separate from each other but deeply interwoven (Stiegler, 1998). However, the increasingly accelerating speeds and scales of technological developments might create a sense of asynchrony, and an illusion of those two having nothing to do with one another. Consequently, the technical is perceived as merely an extension of the biological, an auxiliary instrument that can be universally applied as to transgress natural limitations of human bodies and environments, at any scale and in any place. It is a view that has persistently (mis)informed dominant strains of Western philosophies, protracting a series of discriminatory splits and oppositions, object and subject, nature and culture, or place and space being some of them (Morton, 2018). But there certainly have emerged alternatives for such totalizing myths of technologies as universal instruments of progress, transgression, and solution.[4] Philosopher Yuk Hui proposes a vision in which technologies are to be seen as evolving in an integral, ecological relationship to their locality – its environmental and cosmological conditions – while simultaneously taking into consideration the bigger picture (Hui, 2020). Considering art as integral to one's life (and hence countering yet another split instigated by modernity), Hui suggests that "what needs to be reinvented is not a specific technology that could be more ecological or efficient, but rather a new way of thinking technology in its totality and in its diversity" (ibid., p. 112). Echoing this view, expressed also by Benjamin H. Bratton who sees role of technologies not in augmenting human capacities but in their epistemological function to disclose the world to us in particular ways,[5] what kind of implications can such rethinking have today on soundwalking, approached as not either or, but at once an aesthetic, techno-cultural, environmental, and existential practice? What would it reveal to rethink soundwalking as at once highly subjective, locally conditioned, and yet reliant on and responsive to much larger, planetary relations? Paraphrasing Anna Lowenhaupt Tsing and others' concerns raised in their recent exploration of the arts of living on a damaged planet (Tsing et al., 2017), how can "the tools of modernity be repurposed against the terror of progress to make visible," and I would add *audible*, "the worlds that modernity has ignored and damaged?" (ibid., p. 7). Lastly, can soundwalking become an onto-epistemological technique for expanding and shifting our worldviews from singular, unifying visions of the world (as in *World* Soundscape Project?) towards the plurality of worlds and worldviews out there, those previously unnoticed, systemically ignored, or unintendedly overlooked? These have been yet other important questions that, in synchrony with how art-led research is often pursued, this book intends not to answer but rather bring attention to and reflectively explore.

Undoubtedly, many contemporary artists, including the contributors to this book, would agree that technologies will not solve problems and fix our damaged realities. Instead, they might become channels for generating new or regenerating old, ignored, and silenced techniques and ways of perceiving the world. Echoing Bernard Stiegler, critical and political negotiation of our position within our techno-cultural realms is not about resisting or adapting. Instead, it is about *inventing* (2003, p. 72).[6] Approached from that angle, technologies do not provide any clearer and more correct access to the surrounding environment than do our "bare" senses. They do not merely distort or diminish our perception of the world either. Instead, I suggest, in shifting our focus towards specific registers, scales, subjects, and temporalities, otherwise imperceptible, they might help invent and offer *different* kinds of itineraries along which our human and other than human realms can be experienced. And, perhaps, only after that, differently approached, cared for, and repaired. The same artists might also agree that while technologies immerse their users in those untapped, often difficult, narratives and perceptions of the world, they always do it from a specific angle. This is not only because of biases that technologies and cultural techniques are inherently ingrained with (Srinivasan, 2013; Siegert, 2015; Benjamin, 2020; Lagerkvist, Smolicki, Tudor, forthcoming), or habitual uses thereof that implicitly exclude other forms of experience (Chun, 2016). It is also because of positions from which an artist, activist, and listener acts, listens, and wants others to do so. Thus, no soundwalk is innocent. Every act of walking and listening, no matter how open and inclusive, technologically complex or reduced, collective or individual, spontaneous, or planned, is always to some extent oriented and, in turn, orienting (Ahmed, 2006). During my conversation with Hildegard Westerkamp in the summer of 2020, I asked whether her definition of soundwalking has transformed over the years. The essence, she said, has not changed: soundwalking has been and still is about finding an opportunity to go out and listen to the environment. However, as she added towards the end of her reflection, it is also about "discovering one's own listening." This double orientation of the walking and listening subject, outwards into the environment and inwards as to recognize one's positionality, although certainly present and explored in earlier discussions on soundwalking, seems to be gaining particular momentum today.

This book brings voices of walking and listening practitioners situated and working in and with specific contexts and geographies. By implication, the authors are specifically oriented towards them and, recursively, from them, outwards. Deliberate recognition and acknowledgment of this situatedness and the unparalleled character of itineraries along which they attend to walking and listening as techniques of knowledge making are what constitute departure points for reflective itineraries we are invited to follow. In making room for this kind of oriented (and orienting) practice-based knowledge, this book hopes to push sound-oriented scholarship further away from approaches and practices in which sounds and soundscapes might be a target to rigid categorization, totalizing views (god-tricks, if you will, after Haraway [1988]), managerial control, and discipline.

By bringing researchers and practitioners who personally engage with listening and walking, although with different degrees of familiarity with the concept of soundwalking, the overarching intention of this book is to further expand the ways we – artists, scholars, participants – think, problematize, and practice it. Due to diverse backgrounds of the invited contributors, this revision invites a multiplicity of perspectives that, besides acoustic ecology and soundscape studies, span such fields and areas of interests as Anthropocene, Indigenous studies, history, urban planning, post-humanities, post-colonialism, African American studies, environmental humanities, performance, media arts and design, media studies, political science, migration studies, psychology, poetry, and more. A quick inspection of this still incomplete list of fields that you, the reader of this book will soon be guided through, is enough of a proof that soundwalking needs to be recognized as at once a highly situated and yet exceedingly relational and emergent activity, forever evading containment, and definition. A profoundly inter- or post-disciplinary, centerless endeavor. An *ecotonal* practice (McCartney, 2016) that similarly to environmental ecotones – transitory areas of wetlands, tidal zones, forest fringes, estuaries – cannot be reduced to one site, temporality, discipline, discourse, or worldview. Instead, soundwalking, as offered in this book, persistently leaks into and links diverse perspectives, disrupts old and establishes new fields of tension and collaboration.

This relationality is also a reason for the structure of this book, or, rather, lack thereof. Initially, the book was planned to unfold across three thematic sections, each putting particular emphasis on one dimension of the soundwalking practice: space, time, and technology. However, in the editorial process, it quickly became obvious that structuring the book in such a way would not do justice to the contributed pieces, each, in its own terms and to various extents, engaged with those aspects as always already interwoven realms, often as such in need for redefinition. I believe that letting the chapters simply follow one another, without any rigid boundaries in between (or a red thread facilitating their succession), conduces to a more productive kind of experience, less regimented and hence more open to associations at the reader's end.

The book begins with Usue Ruiz Arana's proposition of soundwalking as a technique for reconfiguring multispecies relations in an urban context (Chapter 1).

Arguing against wildness as something alien to human realms and cultures, the author calls for a revision of how we perceive ourselves in ecological terms. Guiding us through concrete encounters, the author consistently inspires us to think how direct, embodied engagement with other species can contradict the visions of city planners and resulting biases and power imbalances that consequently pervade our urban fabric. Ever intensifying in terms of their number and yet remaining unanticipated, also due to climate change, those encounters require a completely different kind of attitude from that of multispecies relations today. This process, as the author demonstrates, can be facilitated by the practice of soundwalking, presented specifically as a form of *wilding* ourselves and our senses.

In the Chapter 2, Louis Chude-Sokei demonstrates how walking and listening, among many other sonic practices taking place at the intersection of public and private realms, have been and continuously are subject to racial and social stratification. The chapter begins with a reference to Ralph Ellison's *High Fidelity*, a text which, as the author emphasizes, should have long been a foundational one for sound studies. Consequently, Ellison's text becomes a gateway through which Chude-Sokei invites us into soundscapes of Jamaica, a country where he grew up, along with other sites he walked through afterwards. Like the apartment building in Ellison's book, in any racially segregated spaces, soundscapes become sites of competition, resistance, and violence. But also exploration, performativity, and creativity. To understand the social significance of those stratified soundscapes, it is important to turn to the material infrastructure through which they are commenced and sustained: outdoor sound systems, sub-woofers, turn tables, boomboxes, filters and effects, all inseparable from an affective and political experience they convey. It is an experience in which sound does not necessarily absolve or separate, but, in the author's words, *stains*, by forcing identification with the Other.

In Chapter 3, Elena Biserna invites us to reimagine and critically reassess borders by focusing on their soundscapes as experienced from a perspective of a walking body. To aid us in this task, she proposes the term *SoundBorderscape*. The concept attends to walking and listening as tools for questioning the perception of borders as territorially fixed and stable. By bringing us to Hong Kong, one of the sites of her practical workshops, she demonstrates how listening while walking, possibly more than any other way of connecting with the surrounding world, allows us to perceive the border as a constantly shape-shifting *membrane* comprising a multitude of practices, technologies, events, and relations. Conversely, to be able to give an account of borders and their bordering practices, one needs to be able to be open to and align with the multiplicity of positions from which they are experienced and actualized. This means an ability to listen to those narratives, despite them drastically revealing our privileges and biases.

In Chapter 4, Patrick Farmer encourages us to abandon our preconceptions about listening and the self. Through an evocative journey that interweaves his own hearing condition with microbiology, psychology, evolutionary science, and poetry, among other realms, Farmer demonstrates that the acts of listening and walking do not necessarily constitute the self but can also make it fall apart.

Through the author's deeply analytical and yet poetic exploration of the condition of semicircular canal dehiscence he has been suffering and learning from since childhood, we discover that walking and listening are never entirely performed and controlled by a singular agency located and radiating from an individual human being. Instead, they happen due to the assemblage of cooperative processes that meet and recombine in the labyrinth of one's inner ear and its balance organ. Given that soundwalking has predominantly been pursued as an activity in which one grounds oneself in place through intensified sonic attention, Farmer's experience invites us to perceive it as all but that. With a helpful concept of *vertego*, we are encouraged to degravitize this perspective by approaching the inner ear as the very site of soundwalking, a centerless labyrinth through which environments, and their soundscapes, persistently walk.

In Chapter 5, Milena Droumeva begins from a series of timely questions related to artistic approaches to climate change. What does it mean to sonically witness the planetary extinction that unfolds as we speak? What kinds of actions are possible and particularly needed to evacuate listening, field recording, and soundwalking practices from dominant functions they have been comfortably serving, which is to appreciate environments? Lastly, and perhaps most importantly, if we agree that soundwalking and other reflective practices cannot directly combat the effects of human-caused climate change, how can we soundwalk *in* climate change? As one can imagine, and as Droumeva demonstrates, while there are no easy answers to this question, there certainly are indications for how the very foundational concerns that guide those practices can be, and need to be, recalibrated. As a scholar directly concerned with methodological potentials and shortcomings of soundwalking, also due to their affiliation with the former WSP research environment, Droumeva offers a walk-through of some historical and contemporary soundwalks, hinting at *becoming conscious of the unconscious* in the act of listening as the underlying modus operandi of the practice, particularly among more recent approaches. Consequently, Droumeva introduces us to three accounts from their soundwalking exercises in three distinct ecosystems. In conclusion, Droumeva argues for soundwalking that, on the one hand, acknowledges one's positionality and, on the other, recognizes historical legacies and urgencies of today. In doing so, it opens towards an experience of a place significantly different from how one used to know it, more complex and in need for reaction, not appreciation.

In the following, Chapter 6, Amanda Gutiérrez attempts to reframe soundwalking as a decolonial practice. To do so, she puts forth a situated approach to walking, which is deeply informed by her mestizo ancestry, intergenerational storytelling traditions, and immigration experiences. Simultaneously, the author establishes links with the works of others, particularly Gloria Anzaldúa's literary work on critical race theory. Through this interdisciplinary engagement, Gutiérrez arrives at the concept of aural border-crossing. The concept is offered as a specific soundwalking method in which indigenous, migratory, and settler experiences meet and inform the author's incentive to decolonize the ways we construct identity in relation to our lived environments.

In Chapter 7, Tim Shaw offers a set of reflections stemming from his long-term commitment to technologically mediated practices of soundwalking. The main arguments along which Shaw guides us through his works stem from a critique of approaches in which sound(walking) artists separate making and presentation processes. In mainstream approaches to technologically mediated soundwalks, as the author continues, the use of technologies is often instrumental. What determines this instrumentalization is an anticipated effect that the artist intends to project upon their audience. Such soundwalks tend to immerse their audience in a specifically constructed environment, which, paradoxically, shelters the participant from the surrounding environment, particularly when it comes to chance encounters and random forces, inherent to any place. One other risk that Shaw identifies among such curated soundwalks is their authors' ambition to make technologies imperceptible, including technical procedures and unintended artefacts they often generate. They are kept deliberately hidden beyond the horizon of the participant's attention. In contrast to these approaches, Shaw proposes a much more inclusive approach to soundwalking wherein technologies, and a wide array of unanticipated sonic by-products that their unscripted use in public spaces produces, are recognized as constituents of the event, and hence the environment. Through the reflective exposition of Shaw's work, we learn that to account for the environment as a realm of unreconciled and increasingly complex relations, a similar approach needs to be adopted in how one works with the very technologies and techniques for inspiring others to listen.

In Chapter 8 of the book, Paola Cossermelli Messina turns to the notions of Land, sense of belonging, and privilege as ones overlooked in discussions concerned with soundwalking. Drawing on reflections sparked upon revisiting São Paulo, her hometown, the author introduces the reader to the unsettling aura of its current and historical soundscapes. Her sensory experience of the city and its sonic memory has been and continuously is mediated by segregation and exclusion. Traversing specific locations, which in her childhood and following years of youth one would often fear to walk through, Cossermelli Messina attempts to peel off their sonic layers in search for their oppressive foundations. Her series of soundwalks becomes at once an examination of intersecting positionalities and privileges she recognizes within herself, and an attempt at shifting the practice's critical potential towards decolonizing purposes. Consequently, through the author's text, we are invited to engage in soundwalking as a critical tool for destabilizing one's sense of identification with a place and belonging to it, or, even a tool for contesting the very notion of citizenship as a dominant way and right to belong. By inviting the notion of Land that points towards underlying principles, philosophies, and ontologies of space, soundwalking can aid us in reinventing ourselves, in and through the relation to Land.

A starting point for Katt Hernandez, the author of Chapter 9, is a tension that she recognizes between dominant aesthetic approaches to sensing our surrounding environments while walking: psychogeography and soundwalking. Consequently, after introducing us to the histories of those practices, Hernandez sets off on a task to chart what she calls a third way in thinking about sensory exploration of urban environments. Through a purposefully eclectic and non-linear mosaic of memory recollections, anecdotes, and references to works of others, Hernandez questions earlier disciplinary ramifications of the practices in question, which, as she argues in the concluding section, should not seek any boundaries, names, and definitions beyond one's individual experience. Hernandez's chapter can be read as an exercise in an anti-disciplinary and anti-disciplining, which is to say non-discriminatory approach to listening while walking. It is a reminder that by attempting to define and contain what these practices have an unquestionable potential to open us up to, we might often do more harm than good to how we account for places and diversity of experiences they are conducive to.

In the Chapter 10, through a profound engagement with sonic memories of Armenian Genocide victims, Gascia Ouzounian examines the role of ear-witnessing in the construction of individual and collective memory. During several years of systemic persecution conducted by the Ottoman Empire (present-day Turkey), 1.5 million ethnic Armenians were murdered. Her study does not only seek to give justice through hearing to the victims' voices, but to compensate for what visual accounts, predominant as the author asserts, might have omitted. Instead, the kind of listening she is committed to is deliberately expanded in how it traverses diverse aural position in relation to the selected accounts. These positions include awareness of what the victims of the genocide heard at the time, how they felt, responded to, and consequently remembered what they heard. Many of the accounts the author refers to stem from memories of those who, after being expelled from their hometowns, were forced to march to deserts, where many of them perished. Needless to say is that the combination of walking and listening in this specific historical context, during those marches, had completely different meaning and roles from soundwalking in other contexts as discussed on other pages of this book. Ouzounian's diligent engagement with the echoes of this tragic, unreconciled history – a form of soundwalking through its echoes – can hence be seen as a way of expanding our understanding of the practice in question. The chapter offers soundwalking as a technique for carefully returning to and giving justice to difficult events of history through sonic accounts left behind by their victims.

Lastly, in Chapter 11, I, Jacek Smolicki, share insights from my recently developed soundwalk compositions. I adopt the term *soundwalk composition* to suggest that walking and listening are always implicated in and hence entail some form of composition, recomposition, and decomposition. In other words, when soundwalking, we are traversing in part composed territories and soundscapes. These, to unfold as they do and sustain themselves, necessitate(d) decomposition of other environmental, geological, and cultural realms and, consequently, their soundscapes. In this chapter I suggest that what we soundwalk through is not merely the *here* and *now*, but much wider a spatio-temporal territory that spans the past (even deep, geological time) and extends into the unknown future(s). Combining diverse technological and conceptual modes of field recording, mediating, and broadcasting both precomposed material and sounds from the immediate vicinity, soundwalking compositions presented in this chapter bring attention to specific locations: Vancouver's intertidal zones and Swedish most and least polluted lakes. While opening to imperceptible aspects of those geographies, these pieces are simultaneously exercises in (re-)composing one's ethical position within a mesh of scales, agencies, and temporalities that constitute each and every site.

Standing at the square near Vancouver's Roundhouse and listening to the introduction to the soundwalk I will soon join, I think of these untapped and problematic dimensions of soundwalking that are only now being acknowledged and which keep calling for even deeper attention. What prompts my thoughts is the length of the introduction preceding the soundwalk, taking a good half an hour. This contextualization of the event includes: a brief history of the practice; acknowledgment of the unceded territories of Coast Salish people – the xʷməθkʷəy̓əm (Musqueam), Skwxwú7mesh (Squamish), and Səl̓ílwətaʔ/Selilwitulh (Tsleil-Waututh) Nations – on which we already stand and will soon soundwalk through; recognition of capitalist forces generative of economic injustice and problematic developments within the parts of the city through which the walk will unfold; instructions for how the very walk is to be enacted; practical hints on how to listen; and, lastly, assurance of everyone's safety and, in case of injuries, the presence of an assistant equipped with a first aid kit. Dividing my attention between this lengthy introduction and my increasingly impatient daughter, I am made aware of the difference between times when soundwalking was commenced as a soundscape studies method and today. It seems that with every year passed, soundwalking has become less innocent while ever more in need for contextualization, and confrontation with difficult questions. This is, with no doubt, necessary: this openness and continuously emergent character is what arguably allowed the practice to stay afloat, to remain timely and ethically responsive to necessary revisions of history, challenges of today, and visions of the future. However, considering one of the biggest strengths of soundwalking being its simplicity, I wonder to what extent shall we keep complicating it? Where are the limits to preparatory and contextualizing gestures that, at the end of the day, serve to justify the simplicity of a soundwalking event? Does not one risk ending up standing still, earwitnessing an ever more intense pool of concerns that impede any further steps? In alliance with the contributors to this book, I suggest that, instead of treating them as obstacles, we can use such timely concerns as deliberate departure points towards difficult, challenging, and yet rewarding soundwalking itineraries for aurally connecting with multiple scales of our troubled contemporaneity.

Urged by my daughter and her preference of playground noises over the city din, I leave the soundwalk halfway into it. My soundwalking, however, does not end. It only begins. Again and again. After all, every day there are *worlds* listening days.

Jacek Smolicki
Stockholm, Vancouver, July–August 2022

Notes

1 World Listening Day (WLD) was established on July 18, 2010, the birthday of R. Murray Schafer, Canadian composer and the founder of World Soundscape Project, who passed away on August 14, 2021. The aim of this initiative has been to "celebrate the practice of listening as it relates to the world around us, environmental awareness, and acoustic ecology." (After press release available on https://www.worldlisteningproject.org/world-listening-day/#, accessed April 8, 2022.) The soundwalk on July 17, 2022 was led by Jorma Kujala.

2 Consider a number of initiatives, emerging during the pandemic and aimed at capturing changes to acoustic environments in diverse locations. For example. #StayHomeSounds project by Cities and Memories, https://citiesandmemory.com/covid19-sounds/, accessed July 15, 2022.

3 As John Drever reminds us, there is a significant difference between how Hildegard Westerkamp and R. Murray Schafer regarded soundwalking. For Westerkamp, the practice was all encompassing in terms of modalities of listening one invites to the practice. For Schafer, soundwalks differed significantly from listening walks. Listening walks were simple walks with a focus on listening whereas soundwalks were often carefully choreographed and facilitated by scores, maps, and other technologies (Schafer, 1974, p. 17, in Drever, 2020).

4 For example, along with the field of STS (Science, Technology, and Society) the post-phenomenological views of Don Ihde (1979, 1983), Peter-Paul Verbeek (2011), along with the emerging Existential Media perspective pursued by Amanda Lagerkvist (2022) put stress on how the role and agency of technologies is always subject to negotiation (ethical, moral, existential, etc.) within concrete everyday life-worlds.

5 Source: https://palladiummag.com/2021/01/11/benjamin-h-bratton-on-terraforming-the-world-order/, accessed: July 20, 2022.

6 Etymology of the term inventiveness links it to inventory. Thus, inventiveness might mean not so much an inauguration of something new, novel, and original, but re-configuration of components within an already pre-existing (technical) systems. It is about re-thinking connections between existing elements of the world to conceive of a new relationship between them, a new constellation which, though contained within the horizon of possibilities established by existing technical systems, moves beyond their constraints or recombines these constraints so they become productive towards other directions, aims, and motivations (Smolicki, 2017, p. 23).

References

Ahmed, S. (2006) Queer Phenomenology. *Orientations, Objects, Others.* Durham, NC: Duke University Press.

Akiyama, M. (2015) Unsettling the World Soundscape Project. Soundscapes of Canada and the Politics of Self Recognition, Sounding Out sound studies blog, August 20, 2015, https://soundstudiesblog.com/2015/08/20/unsettling-the-world-soundscape-project-soundscapes-of-canada-and-the-politics-of-self-recognition/, accessed: July 12, 2022.

Atkinson, R. (2007) Ecology of Sound: The Sonic Order of Urban Space, *Urban Studies* 44(10): 1905–1917.

Augoyard, J.F., Torgue, H., eds. (2006) *Sonic Experience: A Guide to Everyday Sounds*, trans. A. McCartney. Montreal: McGill-Queens University Press.

Benjamin, R. (2020) *Race After Technology. Abolitionist Tools for the New Jim Code.* Cambridge: Polity.

Chun, W.H.K. (2016) *Updating to Remain the Same. Habitual New Media.* Cambridge, MA: MIT Press.

Drever, J.L. (2020) Listening as Methodological Tool: Sounding Soundwalking Methods. In: M. Bull, M. Cobussen, eds. *The Bloomsbury Handbook of Sonic Methodologies*. London: Bloomsbury Academic, pp. 599–613.

Drobnick, J. ed. (2004) *Aural Cultures*. Toronto: XYZ Books.

Droumeva, M., Jordan, R. eds. (2019) *Sound, Media, Ecology*. New York, NY: Palgrave Macmillan.

Eidsheim, N.S. (2019) *The Race of Sound: Listening, Timbre, and Vocality in African American Music*. Durham, NC: Duke University Press.

Ferrarini, L., Scaldaferri, N. (2020) *Sonic Ethnography. Identity, Heritage, and Creative Research Practice in Basilicata, Southern Italy*. Manchester: Manchester University Press, Openhive.

Haraway, D. (1988) Situated Knowledges: The Science Question in Feminism and the Privilege of Partial Perspective. *Feminist Studies*, 14(3), 575–599.

Hui, Y. (2020) *Art and Cosmotechnics*. Minneapolis, MN: University of Minnesota Press.

Ihde, D. (1979) *Technics and Praxis*. Dordrecht: Reidel.

Ihde, D. (1983) *Existential Technics*. Albany, NY: SUNY Press.

Järviluoma, H., M. Kytö, H. Uimonen, N. Vikman (2009) Soundscapes in Change – From 1975 to 2000. In: H. Järviluoma, M. Kytö, B. Truax, H. Uimonen, N. Vikman eds. *Acoustic Environments in Chance & Five Village Soundscapes*. Tampere & Joensuu: TAMK & University of Joensuu, pp. 224–281.

Kanngieser, A. (2015) Geopolitics and the Anthropocene: Five Propositions for Sound. *GeoHumanities* 1, 80–85.

Kanngieser, A., Todd, Z. (2020) From Environmental Case Study to Environmental Kin Study. *History and Theory* 59(3), Wesleyan University.

Kelly, C. ed. (2011) *Sound*. Whitechapel: Documents of Contemporary Art, Cambridge, MA: MIT Press.

Kytö, M. (2019) Sound Transformations in Space. In: M. Bull, ed. *The Routledge Companion to Sound Studies*. London: Routledge, pp. 177–184.

Kytö, M. (2019a) Ecocriticism and the Anthropocene in the Water Soundscape Composition Contest. In: T. Howell, R. Powell eds. *The Nature of Nordic Music*. London: Routledge, pp. 159–172.

Lagerkvist, A. (2022) *Existential Media. A Media Theory of the Limit Situation*. Oxford: Oxford University Press.

Lagerkvist, A., Smolicki, J., Tudor, M. (forthcoming) Sonorous Surfaces, Biased Backends: The Gendered Voices of AI Assistants as Existential Media. In L. Parks, J. Velkova, S. De Ridder, eds. *Media Backends: Digital Infrastructures and Sociotechnical Relations*. Chicago, IL: University of Chicago Press.

Lane, C. and Carlyle, A. eds. (2021) *Sound Arts Now*. Axminster: Uniformbooks.

Losiak, R., Wieczorek, S. (2021) Ekologia Dźwiękowa. Glissando Audio Papers, source: http://audiopapers.glissando.pl/category/ekologia/, accessed: July 12, 2022.

Loveless, S. (2020) Tactical Soundwalking in the City. A Feminist Turn from Eye to Ear *Leonardo Music Journal*, I. 30, 99–103.

McCartney, A. (2000) *Sounding Places with Hildegard Westerkamp*, Electronic Music Foundation Institute, Doctoral dissertation.

McCartney, A. (2014) Soundwalking: Creating Moving Environmental Sound Narratives, *Oxford Handbook of Mobile Music*, 2, 212–237.

McCartney, A. (2016) Ethical Questions about Working with Soundscapes. *Organised Sound*, 21(2), 160–165.

Morton, T. (2018) *Dark Ecology. For a Logic of Future Coexistence*. New York, NY: Columbia University Press.

Ochoa Gautier, A.M. (2014) *Aurality: Listening and Knowledge in Nineteenth-Century Colombia*. Durham, NC: Duke University Press.

Radicchi, A. (2017) *A Pocket Guide to Soundwalking. Some Introductory Notes on Its Origin, Established Methods and Four Experimental Variations, Atmosphere and Perception.* Berlin: Universitätsverlag der TU Berlin, pp. 70–73.

Robinson, D. (2020) *Hungry Listening. Resonant Theory for Indigenous Sound Studies.* Minneapolis, MN: Minnesota University Press.

Ruiz Arana, U. (2019) Thinking with My Ears – Guidance on Sound for Landscape Architects. In: *Internoise - The 48th International Congress and Exhibition on Noise Control Engineering.* Madrid: i-ince.

Schafer, R.M. (1977) *Soundscape, Our Sonic Environment and the Tuning of the World.* Rochester, VT: Destiny Books.

Siegert, B. (2015) *Cultural Techniques: Grids, Filters, Doors, and Other Articulations of the Real.* New York, NY: Fordham University Press.

Smolicki, J. (2017) Para-archives. Rethinking Personal Archiving Practices in the Times of Capture Culture, Doctoral dissertation, Malmö University, source: http://www.diva-portal.org/smash/record.jsf?pid=diva2%3A1404308&dswid=-7896, accessed: August 12, 2022.

Smolicki, J. (2021) Minuting. Rethinking the Ordinary Through the Ritual of Transversal Listening. *VIS Nordic Journal for Artistic Research*, I. 5, Source: https://www.en.visjournal.nu/minuting-rethinking-the-ordinary-through-the-ritual-of-transversal-listening/, accessed: July 18, 2022.

Srinivasan, R. (2013) Re-thinking the Cultural Codes of New Media: The Question Concerning Ontology. *New Media Society*, 15(2).

Steingo, G., Sykes, J. eds. (2019) *Remapping Sound Studies.* Durham, NC: Duke University Press.

Sterne, J. (2012) *Sound Studies Reader.* New York, NY: Routledge.

Sterne, J. (2019) Multimodal Scholarship in World Soundscape Project Composition: Toward a Different Media-Theoretical Legacy (Or: The WSP as OG DH). In: M. Droumeva, R. Jordan, eds., *Sound, Media, Ecology.* New York: NY: Palgrave Macmillan, pp. 85–109.

Stiegler, B. (1998) *Technics and Time, 1: The Fault of Epimetheus.* Stanford, CA: Stanford University Press.

Stiegler, B. (2003) Our Ailing Educational Institutions. *Culture Machine*, 5, Source: https://culturemachine.net/the-e-issue/our-ailing-educational-institutions/, accessed: July 19, 2022.

Truax, B. (1974) Soundscape Studies: An Introduction to the World Soundscape Project. *Numus West*, 5.

Truax, B. (1999) *Handbook for Acoustic Ecology* [CD-ROM]. Burnaby, BC: Cambridge Street Publishing, www.sfu.ca/~truax/handbook.html, accessed: August 19, 2022.

Truax, B. (2012) Sound, Listening and Place: The Aesthetic Dilemma. *Organized Sound*, 17(3), 193–201.

Truax, B. (2019) Acoustic Ecology and the World Soundscape Project. In: M. Droumeva, R. Jordan, eds. *Sound, Media, Ecology.* New York, NY: Palgrave Macmillan, pp: 21–44.

Tsing, A.L., Swanson, H.A., Gan, E., Bubandt, N., eds. (2017) *Arts of Living on a Damaged Planet. Ghosts and Monsters of the Anthropocene.* Minnesota, MN: University of Minnesota Press.

Verbeek. P.P. (2011) *Moralizing Technology: Understanding and Designing the Morality of Things.* Chicago, IL: University of Chicago Press.

Westerkamp, H. (1974) Soundwalking. *Sound Heritage*, 3(4), updated in Angus Carlyle, ed. *Autumn Leaves: Sound and the Environment in Artistic Practice*. Paris, 2007, pp. 49–54.

Westerkamp, H. (1994) The Soundscape on Radio. In: D. Augaitis, D. Lander, eds. *Radio Rethink: Art, Sound and Transmission*, Walter Phillips Gallery, The Banff Centre for the Arts.

Wright, M.P. (2022) *Listening After Nature. Field Recording, Ecology, Critical Practice*. London: Bloomsbury.

Zylinska, J. (2014) *Minimal Ethics for Anthropocene*. London: Open Humanities Press.

1

SOUNDWALKING IN THE PHONOCENE: WALKING, LISTENING, WILDING

Usue Ruiz Arana

Being prey

In the Australian winter of 1985, ecofeminist Val Plumwood survived a saltwater crocodile attack whilst canoeing along the paperbark wetlands of Kakadu National Park. In the aptly named paper 'Being Prey', Plumwood captured the encounter with the crocodile (1996). Holding each other's gazes in the rain and wind. The subsequent attack of the canoe. Her failure to consult the indigenous Gagagdu owners of Kakadu before her visit. Her attempt to climb into a paperbark tree. The realisation, lastly, of being food for others as the crocodile grabbed her between the legs and drove her into the water multiple times. In this struggle, Plumwood gained a glimpse of the world 'from the outside' (ibid., p. 34); a world where she was no longer a subject but an object to satisfy others' needs. This newfound world was indifferent to her death, and would, therefore, carry on without her. Plumwood acknowledged this vulnerability as necessary to 'realize how misguided we are to view ourselves as masters of a tamed and malleable nature' (ibid., p. 44).

Thirty years on, Plumwood's call to expand beyond an instrumental view of nature is more urgent than ever as we race against time to tackle our climate and biodiversity emergency. In our climate and biodiversity emergency, Nature-Based Solutions (NBS), interventions that draw from natural processes, have become an integral part of urban design discourse and planning. However, they have yet to deliver the needed outcomes at the required speed (Randrup et al., 2020). To expand beyond the instrumental regard for nature inherent to NBS, the notion of Nature-Based Thinking (NBT) has been proposed, to move away from anthropocentrically designed and managed urban and rural places (ibid.). In the context of both NBS and NBT, the word nature is employed, grammatically, to denote the non-human, and it is this connotation that is retained in this chapter.

DOI: 10.4324/9781003193135-2

The soundwalking practice advanced in these pages is a tool that can be encompassed in Nature-Based Thinking, as it can give us glimpses of the world 'from the outside' (Plumwood, 1996, p. 34) by facilitating encounters with a variety of other beings, and can therefore inform the co-design and co-management of cities with a multitude of agents.

This chapter traces the development of this wilding practice through experiential soundwalks and encounters with non-human beings. The chapter starts by contextualising this approach within Nature-Based Thinking, and by trailing the return of walking as an aesthetic practice for urban design. Then, through reflections on key experiential walks, and drawing on ecofeminist theory and sound artists, this chapter follows the development of my soundwalking practice from its tentative beginnings to a tool for wilding.

Nature-Based Thinking

In Landscape Architecture and related disciplines – mainly urban design and planning – we envisage a greener and wilder future for our cities. Working closely with a variety of other disciplines such as planners, ecologists and engineers, and drawing from a growing body of research that demonstrates the health and well-being benefits of contact with nature, we develop and integrate Nature-Based Solutions (NBS) into our urban schemes to help tackle climate change, increase biodiversity and connect citizens with nature at their doorstep (Tozer and Xie, 2020; van der Jagt *et al.*, 2020; Council of Europe, 2021).

The concept of NBS is the latest in a series of approaches that seek to integrate nature into cities (Randrup et al., 2020). NBS is defined as 'solutions that are inspired and supported by nature … provide environmental, social and economic benefits and help build resilience … through locally adapted, resource-efficient and systemic interventions' (European Commission, 2015, online). In practice, despite their many benefits, NBS retain an instrumental view of nature. The focus is on what nature can do for us – for the development of technical and quantifiable solutions – and on nature as a place to go to. Moreover, NBS are failing to redress the climate and biodiversity emergency at the required rate (IPCC, cited in Randrup et al., 2020). As a response, and to expand NBS, Randrup et al. return to the foundation of NBS as 'inspired by nature' to inform their 'Nature-Based Thinking' (NBT) proposition – a holistic approach to wilder cities founded on the 'interconnectedness of humans, nature and technology' (2020, p. 920).

Nature-Based Thinking (NBT) draws from environmental philosophy, urban geography and non-Western conceptualisations of nature 'to look at the relations between cities and nature in a more cyclical way' and allow for 'culturally diverse' ways of understanding nature (Randrup et al., 2020, p. 921). A key component of NBT is the need for urban citizens to experience the 'unpredictability and heterogeneity of nature', which is key to establish an emotional connection with others (ibid., p. 922). Kinship with non-human others allows us to develop an ethic of care towards them that acknowledges their intrinsic value (Bennett, 2001).

Kinship with others is the focus of my soundwalking, as the reader will discover through these pages, as shifting cultural values underpinning our relationship with nature triggers deeper environmental change than shifting behaviours within retained instrumental values (Ives and Fischer, 2017).

Soundwalking and walking

Soundwalking in urban design and planning currently fulfils different interrelated, human-centred purposes. Soundwalking is central to the Soundscape Approach that draws from the acoustic ecology tradition to evaluate the soundscape from human perception, to interpret and modify said soundscape (BSI, 2014, 2018, 2019).

Soundwalking is also a tool to uncover 'the perceptual relationships between the inhabitants and the city itself through its sonic component' (Radicchi, 2017, p. 70), and as such is included within embodied walking practices that seek to engage all the senses in the critical analysis, interpretation and re-imagining of the landscape (Careri, 2017; Schultz and van Etteger, 2017). Architect Francesco Careri builds a case for reclaiming walking as an aesthetic practice through a historic overview of walking from the Nomads through key twentieth-century walking movements, including the dadaists, surrealists, situationists and land artists (2017). Through this historic analysis, Careri argues that walking has always been a way to read and write the landscape, yet we have lost it in the day-to-day design of our cities. We read the landscape as we traverse through it and perceive it. And we write the landscape through our very presence in it, through our activation of the spaces we traverse (ibid.). The soundwalking practice advanced in this chapter builds on the latter: walking as a vehicle to re-read and re-write the landscape. With a focus on attentive listening, I propose, walking can help us move human-centrism, and align soundwalking with Nature-Based Thinking approaches, as discussed next through three experiential walks and reflections.

Soundwalks

The cries of the children lost at sea (Walk 1, Quayside, Newcastle upon Tyne, UK)

Kitte-wa-aaake, kitte-wa-aake, kitte-wa-aake … kitte-wa-aake, kitte-wa-aake, kitte-wa-aake. I first heard the piercing call of the kittiwakes on a Saturday afternoon along Newcastle's quayside seven years ago. It was the tentative beginning of my soundwalking practice, which at the time consisted simply of spurs of attentive listening incorporated into my daily walks as an aesthetic pursuit.

On that early April afternoon, I went for a leisurely walk with no destination in mind. As I came down Dean Street, a steep Georgian street that follows the dene of the Lort Burn, screams of kittiwakes piqued my curiosity, turning my attention towards Newcastle's Tyne bridge tower. There, hundreds of kittiwakes perched on its edges created a ceiling of sound that mingled with the drunken screams of football fans celebrating a home win in nearby pubs, and the roar of the traffic flowing over the bridge. I continued walking along the quayside guided by the calls of the kittiwakes, crossing over the River Tyne to Gateshead, to finally reach another ceiling of sound: the Baltic, a former flour mill, now a centre for contemporary art and home to a further colony of kittiwakes. Here, the sound of traffic and the drunken screams were a distant murmur and I focused on the kittiwakes. I stood still for some time, observing some taking flight, listening to others perched above, the ceiling of sound no longer a ceiling but a distinctive dialogue of calls. A sudden lamenting cry close to my ear drew a shiver down my spine and I let out a cry of my own, as the kittiwake sounded just like my toddler keeping me awake at night. Aptly, in the local folklore, as I later learnt from field recordist Chris Watson, the cries of the kittiwakes were thought to be the cries of the children lost at sea.

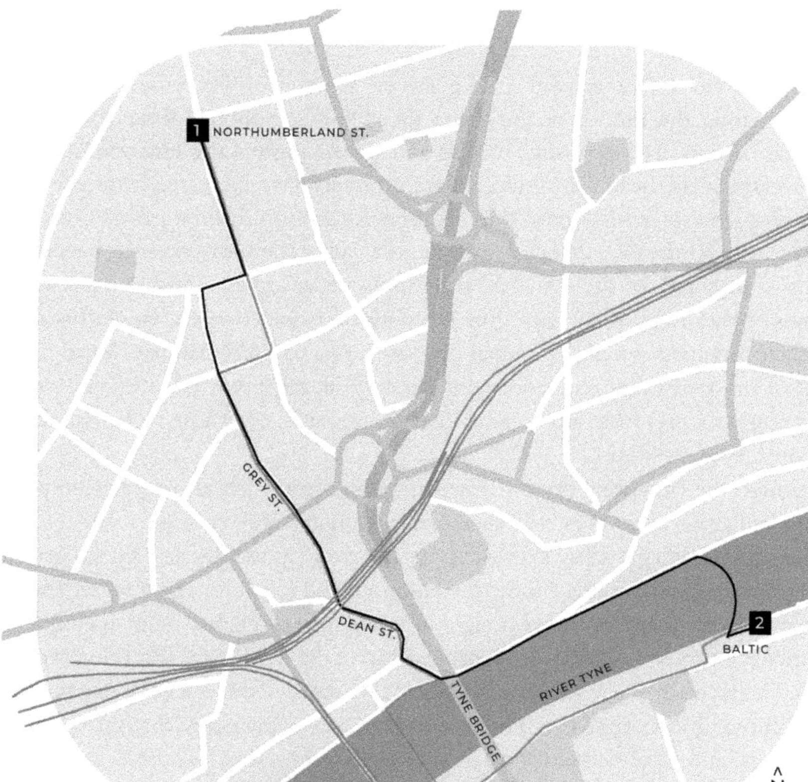

FIGURE 1.1 'Walk 1'. Base map drawn by Usue Ruiz Arana from © OpenStreetMap contributors (openstreetmap.org/copyright) open data, licensed under the Open Data Commons Open Database License (OBbL) by the OpenStreetMap Foundation (OSMF)

That initial walk, the unexpected hubbub encountered and my interaction with it, made me realise how little I listened previously, as I had walked that same walk countless times before without looking up. The main purpose of listening whilst walking, I found and confirmed through countless walks thereafter, was to be present, alert to, and engaged with my surroundings, a finding shared with many others (Westerkamp, 2007; McCartney, 2014). That initial walk, and in particular, the kinship that I felt through the lamenting cry resembling my toddler, also sparked an ongoing interest and love for a seabird that has long bred along the coastline and on key structures of the city's quayside; a migratory seabird that year after year returns from the North Atlantic Sea to inhabit our coastal cliffs and city and reunite with their established partners for the nesting season. Whereas the North Sea colony of kittiwakes has declined by 40 per cent since the 1970s (Fowlie, 2017) due to factors including climate change and the fishing industry that reduce their prey (fish, shrimps and worms), the Newcastle–Gateshead colony is thriving, in part through the weather protection offered by the city (Redfern and Bevan, 2014).

The kittiwakes first arrived at the River Tyne in 1949, when they took buildings along the mouth of the Tyne for cliffs. Demolition of key buildings, alongside nesting deterrents such as nets and spikes, pushed the kittiwakes inland to Newcastle–Gateshead's quayside, where numbers have increased from a few to over a thousand since the 1990s, thus creating the world's furthest inland breeding colony (Tyne Kittiwakes, 2021). For some, the call of the kittiwakes has become a seasonal soundmark of the quayside and it presents an opportunity to experience wildness at the heart of the city. For building users and owners, the kittiwakes, with their messiness, smell and cacophony, are a pest to be kept at bay. Yearly, the arrival of the kittiwakes coincides with news of illegal spikes and nets, and with conflicting responses from ardent advocates on one side, and those who wish them gone on the other (ibid.).

Through the lamenting cry on that April afternoon, I felt an unexpected connection with the kittiwakes and their struggle, but I also became aware of the conflicting feelings that kittiwakes elicit: joy for some, powerlessness for others. In trying to make sense of these conflicting emotions and relations, and how we might come to address them in urban design and planning to enable multi-species environments, I turned to Plumwood's proposition of the 'ecological self' (Plumwood, 1993). The feelings of joy and powerlessness that the kittiwakes evoke exemplifies the kinship and difference that Plumwood proposes is necessary for our relations with others, and to understand ourselves in ecological terms. We need kinship to overcome self-interest and develop care for others based on a desire for them to thrive, and we need distance to acknowledge their needs as different from ours. Thus, in Plumwood's proposition, the ecological self is developed through 'interaction and involvement with the other', to transform and restrict one another (1993, p. 153).

The connection that I felt that day in April, and that others feel every spring with the return of the kittiwakes, drives some of us to raise awareness of kittiwakes, make space for them and acknowledge them as seasonal and iconic co-makers of public space. The difference that others experience, the powerlessness of being unable to stop them nesting and breeding outside their window, gives way to conflict. This powerlessness and conflict, as experienced through soundwalking, is the subject of the next walk.

Staying with the trouble (Walk 2 – Kensington Market, Toronto, Canada)

By the time of this next walk in Toronto, in the spring of 2015, I had formalised my soundwalking approach. Whereas in the beginning my walks were random daily excursions where I was attentively listening, now I was following the late Schafer's advice on using a score as a guide to explore the environment (Schafer, 1977). My scores were mapped routes planned to incorporate a range of environments, habitats, and species along the periphery of the city (Ruiz Arana, 2019). Some of these soundwalks I carried out on my own, some others in pairs and small groups. Despite this formalisation, I had developed a habit of coming in and out of attentive listening in mundane walks, and it was the listening embedded within these walks and the unpredictability experienced within them that affirmed soundwalking as a wilding practice.

This is a walk that I repeated many times with slight variations – from home to my children's pre-school in Toronto's Kensington Market neighbourhood. An early spring morning, once the ice and snow had finally melted, and as we were turning the last corner of our journey, a sudden rustling of leaves alerted me to a presence: a rat. A rat running out of a low shrub before quickly disappearing behind a wheelie bin; scared, perhaps, by the cracking of a small stick as I walked past. Following this first encounter, I saw the rat (or a rat) often in the walk, around the same spot, and I developed a heightened awareness to the rustling of leaves. In Toronto, there were often animal rustlings: a rat, a squirrel, a raccoon. As a newcomer to the city, the raccoons and squirrels were a novelty and a joy to find. However, my encounters with the rats were unnerving and I tried to keep them short, walking away as fast as I could. That is until the rats came to our back garden early summer, and I could no longer walk away.

One morning I was admiring the wildness in our garden whilst I drank a cup of coffee. My coffee turned bitter as I watched a rat push her babies over the boundary fence. Our neighbours had a rat nest in their crumbling shed abutting our garden. Over the next weeks, we saw the rats often, running across the garden, hiding below our deck, day and night. My kids, used to petting rats at pre-school, sought them out in the garden; the rats, wisely, hid away from them. The advice from our neighbours was to get rid of them, to prevent them from finding their way into the house come winter. As the baby rats grew, visions of rats climbing into my children's beds kept me awake at night. After several sleepless nights, I wanted the rats gone, and so did all nearby neighbours, as the rats' playground expanded to include several adjoining back gardens.

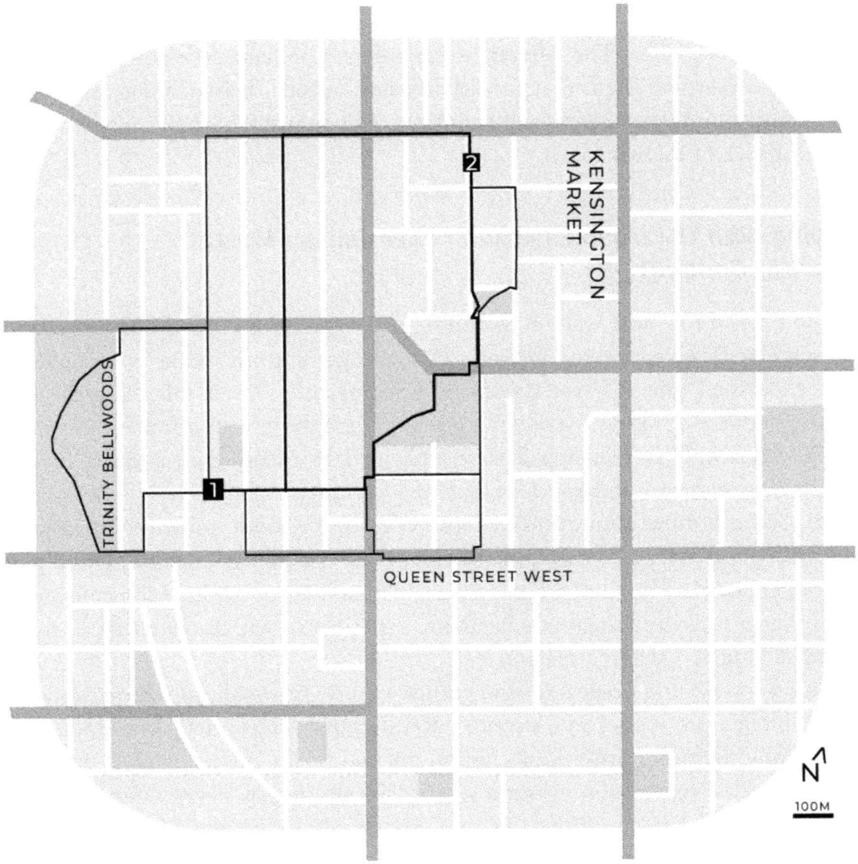

FIGURE 1.2 'Walk 2'. Base map drawn by Usue Ruiz Arana from © OpenStreetMap contributors (openstreetmap.org/copyright) open data, licensed under the Open Data Commons Open Database License (OBbL) by the OpenStreetMap Foundation (OSMF)

Six weeks after the first encounter with the rats, all the neighbours got together and the pest control company that we hired got rid of them. I felt relieved, free from the invasion of what I felt was our territory. And yet, I felt unease, remembering the mother rat guiding her babies, caring for them, the same way that I cared for my own.

Around the time of this encounter with the rats, Donna Haraway was in the midst of writing *Staying with the Trouble: Making Kin in the Chthulucene* (Haraway, 2016). The Chthulucene captures the dynamic relations and forces that we are a part of, and joins the Plantationocene and Capitalocene in troubling the Anthropocene epoch, that for Haraway fails to seize the reality of the times that we are facing, as 'no species, not even our own arrogant one … acts alone' (2015, p. 159). The Chthulucene is a 'timeplace for learning to stay with the trouble of living and dying in response-ability on a damaged earth' (2016, p. 378). Staying

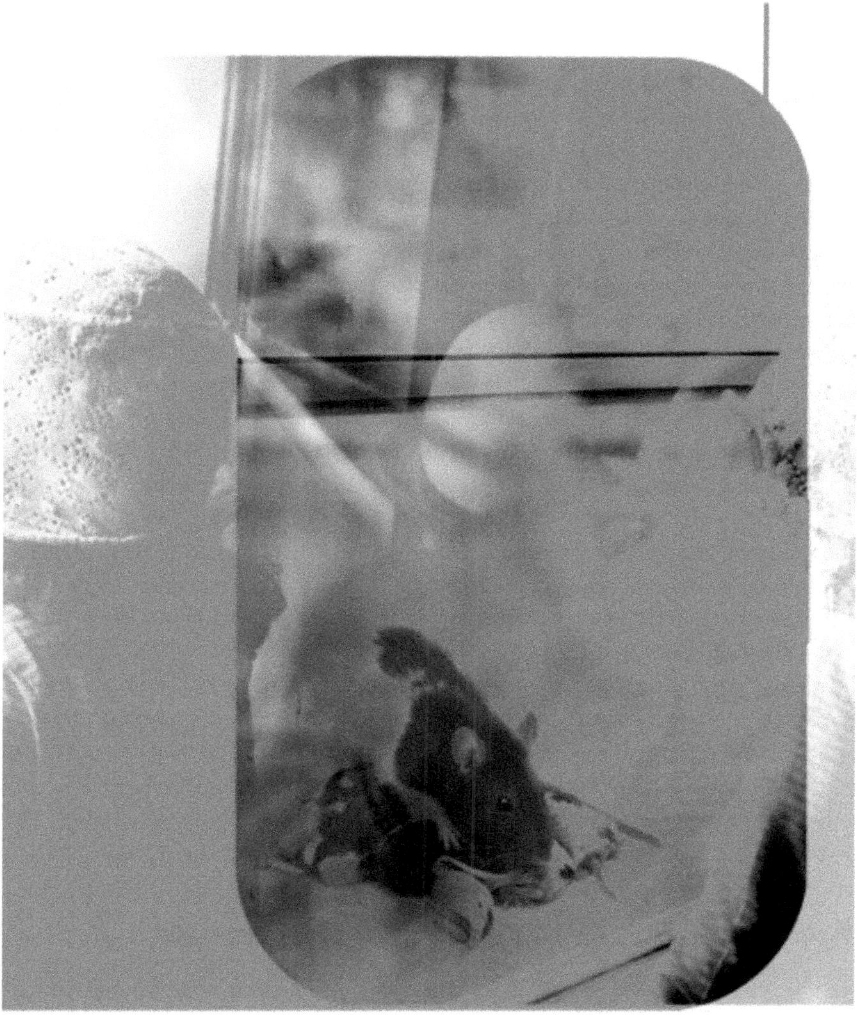

FIGURE 1.3 'Looking out'. Collage © Usue Ruiz Arana. Rat in collage derived from 'Fancy Rat feeding its babies in front of white background' by Eric Isselée, used under Royalty Free Extended license

with the trouble is being present, facing the joys and sorrows that come out of being mortal beings in a multitude of reciprocal configurations and entanglements, and not casting them aside for an imagined ideal future. Aptly, the slogan of the Chthulucene is to 'Make Kin Not Babies!' (2016, p. 102). Making kin and oddkin, not only with our own kind but also across species, is central to staying with the trouble in the understanding that we need each other in a multitude of collaborations to move forward (ibid.)

Unfortunately, Haraway's insistence to make kin and stay with the trouble came too late for us in Toronto. In place of making kin, we cast the rats away, as many other urban creatures are cast away, on an individual basis, house by house, as opposed to strategically, understanding the rats as part of the city's ecosystem (Byers et al., 2019). In doing so, we abstracted the rats from our shared environment, and regarded them as objects to be discarded, disregarding that they are subjects with their own needs and realities (Wolch, 1996). The rats, and other urban opportunistic creatures that we fear as carriers of disease and dirt, seldom feature in idealistic visions of biophilic cities, or in diagrams showcasing the species that our public realm schemes will attract in time. Whereas we think strategically when it comes to protected species and wanted animals, including those that carry disease (such as bats), researching their behaviours, and enabling an interconnected network of habitats to cater for them, we do not plan for the inconvenient, and we do not let it be either.

Wild cities, as multi-species environments, are inevitably sites of interaction, where each creature 'transforms and limits the other' (Plumwood, 1993, p. 156). The resulting entangled and messy environments are inseparable from their constituents, as is the case of the rats and humans that have for centuries thrived alongside one another in settlements. Designing and planning 'from this entangled position demands an acceptance of entanglement without idealisation or despair', as human–nonhuman entanglements are neither fully good nor bad, they simply are (Wright, 2014).

Staying with the trouble is integral to understanding ourselves in ecological terms and moving beyond the discourse of urban nature as something to manipulate from the outside or as a place to go to experience a myriad of health and well-being benefits. The first two walks have developed soundwalking as an interactive and inter-subjective activity that invites us to stay with the trouble. The next and final walk further develops listening as staying with the trouble.

Phonocene (Walk 3, Quayside, Newcastle upon Tyne, UK)

On my return to Newcastle in 2016, and every spring since, I have repeated that initial walk many times, down to the Quayside, seeking the assurance, as I turn the last bend of Dean Street and start to hear the faintest of calls, that the kittiwakes have once again returned.

The walk that I took in 2020, was, however, drastically different as the Covid-19 pandemic took hold. The first lockdown in the UK coincided with the beginning of spring, and the return of the kittiwakes. I ventured down to the Quayside just once, early April, early morning. Unable to use public transport, and no longer working in my office, I started the walk from home, two miles away from my usual starting point. I began walking anxiously and alert, not fully knowing how far from home I could walk as part of my allocated once daily exercise, sensing that perhaps I should not be walking at all. I followed well-trodden streets, once buzzing with people and cars, now deserted. Neighbouring streets resembled parking lots with cars dominating streetscapes, and the city centre resembled a scene from a Sci-Fi movie with its closed shops and absence of human life. That spring morning, as

I took once again the last bend of Dean Street, I heard the familiar call of the kittiwakes. This time, and after the un-realness of the city centre, I was surprised and relieved to find them in their hundreds by the Tyne bridge tower and the Baltic centre for contemporary art. Against the silenced city, the lamenting call of the kittiwakes was more prominent and poignant than ever. Their lament was after all not for us, but despite us. I sensed, in a similar vein to Plumwood's experience of the crocodile that opened this chapter, their indifference to the struggle that we, humans, were facing. The kittiwakes, like many other non-humans in the city, were carrying on without us, occupying the voids left by our absence, thriving.

FIGURE 1.4 'To the Kittiwakes'. Collage © Usue Ruiz Arana

Much has been said and written about the pandemic and its drastic effects on every aspect of our lives. And much has been said and written about its effects on our cities' soundscapes. Cities became quieter, yet noise complaints from neighbours increased (Tong et al., 2021) and the loss of public life was mourned (Ouzounian, 2021). Cities became wilder, as animals seized this newfound silence to venture further into towns (ibid.) and enrich their calls (Derryberry et al., 2020), and researchers took the opportunity to study animal behaviour in unparalleled circumstances (Bates et al., 2020; Rutz et al., 2020).

In a paper for the *Crafting a Sonic Urbanism: Listening to Non-human Life* colloquium, Gascia Ouzounian linked Donna Haraway's proposition of the Phonocene to the pandemic's perceived profusion of birdsong (2021). Two years previously, Haraway, in an interview for *Le Monde*, had hinted at the Phonocene as a companion to the Chthulucene, for the ears 'to hear the sound of… the Earth' and its reciprocal kinships (Vincent, 2019). During a later conversation with Vinciane Despret for the opening event of *Biennal de Pensament*, Haraway termed this listening in kinship as the Phonocene, heard through the human and non-human voices and silences of the disease (2020). For Ouzounian, during the pandemic, the Phonocene was heard not only through the 'sounds of public alerts, ambulances, sirens, disease, and death' but also and 'perhaps counter-intuitively … through the increased presence of birdsong' (2021). Ouzounian linked the dominance of birdsong to the environmental decline that makes zoonotic diseases such as Covid-19 more likely, and that in the case of Covid-19 put a halt to public life and consequent sound making. Similarly, in that walk, I heard the Phonocene through the kittiwakes' cries that punctuated and accentuated human silence.

Human silence was ear-opening as it provided me (and others) with a glimpse of the world 'from the outside' (Plumwood, 1996, p. 35), bringing our ecological vulnerability to the forefront through the realisation that the non-human world was very capable of carrying on without us. Human silence was also an affirmation of the non-humans' place in city-making. That April morning, by the Baltic, the kittiwakes filled our silence with their calls, animating and making the urban environment. And that April morning, in turn, I stayed with the trouble by listening to this broken silence, implicated and entranced by the persistent calls that now demanded all my attention.

Vinciane Despret developed Haraway's proposition of the Phonocene at a presentation for the *Riga International Biennial of Contemporary Art* (Despret, 2020). Despret called for a shift in our understanding of birdsong from an aggressive tool to defend private territory, to an exhibition, a show. She opened her argument with the example of blackbirds' songs that are simultaneously songs to define individual territories and songs learnt from neighbours. To solve this apparent contradiction, and backed by scientific research, Despret invites us to reconsider shared territorial birdsong as a tool for making communities, territories as places for birds to attach to, and territorial borders as meeting places with neighbours. Neighbours, Despret argues, give meaning to territories, and collectively create neighbourhoods. Per Haraway, Despret calls to leave behind the guilt of the Anthropocene and reclaim it as the Phonocene or era of song, to trust the music of the world, including its roars; to enable silences to listen to stories shared by others; and, finally, to follow the emotions that those songs elicit, and find joy and make communities through them (ibid.).

In that April's soundwalk, kittiwakes joined Despret, Haraway and Ouzounian in declaring the Phonocene through the power of their calls, making listening central to how they connected with each other, and how I connected with them. I was listening not to a ceiling of sound, but to a distinctive set of calls, part of the social stimulation these colonies rely on to synchronise and accelerate their breeding (Aubin et al., 2007). Kittiwakes build their nests close together, leaving a gap that enables, through their cries, the meeting of neighbours and the subsequent success of the colony (ibid.). The cries of the kittiwakes are anything but calm, as we have come to expect of birdsong; they are loud, constant, demanding. Yet, staying with the trouble by listening to their thunderous communication showed me, and many others, that there are alternatives to the sonic domination of humans: silences that enable the stories of the non-human makers of the city to be heard.

These three soundwalks have taken us from Newcastle to Toronto and back, and from the 'ecological self' (Plumwood, 1993) to the Phonocene (Haraway in Vincent, 2019; Despret, 2020; Haraway and Despret, 2020; Ouzounian, 2021) through the Chthulucene (Haraway, 2016). The next section develops from these walks, reflections and theories to position soundwalking as a wilding practice and an invaluable tool for the design and planning of our cities in the era of song.

Soundwalking as wilding practice

The introduction situated soundwalking within practices that seek to evaluate and modify the soundscape (BSI, 2014, 2018, 2019), and within broader walking approaches geared towards reading the landscape and writing it through our very presence and activation of it (Careri, 2017). The soundwalking approach described in these pages has sought to build and advance on these human-centred approaches, to position soundwalking as a wilding tool, as part of Nature-Based Thinking approaches that expand beyond the instrumental view of nature inherent to current nature-based solutions to decentre human agency in urban design and planning (Randrup et al., 2020).

The first walk through Newcastle's Quayside introduced soundwalking as a tool for being present and alert to the surroundings, as it was attentive listening that guided my walk, prompting me to follow the call of the kittiwakes and later to find out about their fate and struggle and seek encounters with them for years to come. The first walk was also the beginning of wilding or understanding ourselves in ecological terms, through the feelings, mostly of joy but also discomfort experienced through the encounter with the kittiwakes. Through the walk, it was the kinship experienced through listening to the cry of a kittiwake, the one resembling my toddler, that drove me to connect with them. And it was the distance that I experienced momentarily through the out-of-place cacophony, the smell and messiness, and that others experience as a seasonal struggle, that gave way to conflict. Distance and conflict, however, are required to identify others as different from us, to prevent us from projecting our needs onto them (Plumwood, 1993), and embrace an ethic of 'respect and mutuality, caring and friendship' beneficial to all (Wolch, 1996, p. 27). It is this distance and conflict that I sought to investigate through my second walk.

The second walk through the streets of downtown Toronto developed soundwalking from a tool for being present to an inter-subjective activity. I reacted to another's sound-making – the rustling of leaves; others reacted to my own sound-making – the rat in the street running away as I crunched a branch with my foot; the baby rats hiding from my children as they come out to the backyard to play. The second walk also developed on distance and conflict through the messiness and complexity of those inter-subjective implications, particularly close to home. In Toronto, with the rats, we did not stay with the trouble, as Haraway (2016) advocates; we got rid of it, typical of the way we treat many unwanted creatures. Staying with the trouble requires facing conflict and mess as we make kin with the myriad of others that we are entangled with, as we cannot abstract ourselves or others from the environment that we live in. Staying with the trouble requires once again welcoming commonality and difference and making space for it, as it implies regarding others as subjects with their own worldviews (Wolch, 1996). Staying with the trouble, therefore, requires that we analyse, design and plan our cities not from the outside, from the comfort of our desks, but through and from practices, such as soundwalking, that place us in the inside, in the 'mutual, obligatory, non-optional, you-can't-just-cast-that-away-when-it-gets-inconvenient, enduring relatedness that carries consequences' of our everyday relations (Haraway in Paulson, 2019). Such practices help us to expand site analysis into kin analysis, and design and planning into kin making, in a similar vein to the move from case study to kin study that AM Kanngieser and Zoe Todd call for in environmental histories (2020). Kin analysis and kin making, in turn, enables us to understand site not only as a physical space(s) but more importantly as the stage for a multitude of dynamic, interdependent and complex relations.

The third and final walk laid bare that we are not the only city makers, as it gave us a glimpse of the world 'from the outside' (Plumwood, 1996, p.34) in the recently declared Phonocene or era of song (Haraway in Vincent, 2019). With the return of spring at the start of the Covid-19 pandemic, I heard the Phonocene (as advanced by Ouzounian, 2021) in the vulnerability embedded in human silence, as well as through the cries of the kittiwakes that, like many other non-human creatures, carried on regardless, thriving in the acoustic and physical voids left by our absence, and calling for us to listen. Listening in the Phonocene invites us to stay with the trouble, and to acknowledge our vulnerability in the process, as to listen requires to be open to all sound (Westerkamp, 2015) and consequently, to encounters with unknown worldviews as we attune to the environment (Kanngieser, 2020). Listening whilst walking, or soundwalking is, therefore, an invaluable tool in the Phonocene, as it enables us to relate to others through their songs and silences. Listening to others' stories, invites us, as Despret proposes for territorial birds, to reconceptualise territories of all scales and their borders as inter-species meeting places, as kin making, paving the way for multispecies neighbourhoods.

Wilding in the context of this chapter has become not only the wilding of an urban environment that is outside of us, but more importantly our own wilding as an integral part of that environment. The wilding of our selves – through the realisation that we are implicated in inter-subjective relations with many others and through the recognition of kinship and difference embedded in those entanglements. In the recently claimed Phonocene, soundwalking facilitates inter-subjective encounters, and attunement to other worldviews in kin analysis and for kin making. Whilst we soundwalk, we become landscape – a subject experiencing the spaces traversed and encountering others in the process, as well as an object activating space for other city inhabitants to sense (Careri, 2017). These dialectic encounters, as we have discovered, are sometimes full of joy, other times full of sorrow, or somewhere in between, and both experiences are needed for our own wilding: to understand ourselves in ecological terms and move forward and re-read and re-write cities attuned to the myriad of others that we are currently entangled with. In this manner, soundwalking becomes a wilding practice: the process through which we uncover our wildness as we relate to and consider how to coexist and leave space for a multitude of other beings.

References

Aubin, T., Mathevon, N., Staszewski, V., and Boulinier, T. (2007) 'Acoustic communication in the Kittiwake *Rissa tridactyla*: Potential cues for sexual and individual signatures in long calls', *Polar Biology* 30(8), 1027–1033. doi:10.1007/s00300-007-0262-6

Bates, A.E., Primack, R.B., Moraga, P., and Duarte, C.M. (2020) 'COVID-19 pandemic and associated lockdown as a "Global Human Confinement Experiment" to investigate biodiversity conservation', *Biological Conservation* 248(May), 108665. doi:10.1016/j.biocon.2020.108665

Bennett, J. (2001) *The Enchantment of Modern Life: Attachments, Crossings and Ethics*. Princeton, NJ: Princeton University Press.

BSI (The British Standards Institution) (2014) 'BS ISO 12913-1: 2014 Acoustics – Soundscape Part 1: Definition and conceptual framework'.

BSI (The British Standards Institution) (2018) 'PD ISO / TS 12913-2: 2018 BSI Standards Publication Acoustics – Soundscape', *BSI Standards Publication*.

BSI (The British Standards Institution) (2019) 'PD ISO / TS 12913-3: 2019 BSI Standards Publication Acoustics – Soundscape – Part 3: Data analysis', pp. 1–30.

Byers, K.A., Joseph Lee, M., Patrick, D.M., and Himsworth, C.G. (2019) 'Rats about town: A systematic review of rat movement in urban ecosystems', *Frontiers in Ecology and Evolution* 7(Jan), 1–12. doi:10.3389/fevo.2019.00013

Careri, F. (2017) *Walkscapes: Walking as an Aesthetic Practice*. Ames: Culicidae Press.

Council of Europe (2021) *Urban Landscapes and Climate Change: The Contribution of Landscape Architects to Improve the Quality of Life*. Strasbourg: Council of Europe. Available at: https://rm.coe.int/11th-council-of-europe-conference-on-the-european-landscape-convention/1680a26a86

Derryberry, E.P., Phillips, J.N., Derryberry, G.E., Blum, M.J., and Luther, D. (2020) 'Singing in a silent spring: Birds respond to a half-century soundscape reversion during the COVID-19 shutdown', *Science* 370(6516), 575–579. doi:10.1126/SCIENCE.ABD5777

Despret, V. (2020) 'Phonocene: Bird-singing in a multispecies world', in *Riga International Biennial of Contemporary Art (RIBOCA)*. Riga. Available at: https://www.youtube.com/watch?v=U90M8rhQI6c

European Commission (2015) *Nature-based solutions, European Commission*. Available at: https://ec.europa.eu/info/research-and-innovation/research-area/environment/nature-based-solutions_en (accessed: 8 October 2021).

Fowlie, M. (2017) *Kittiwake joins the red list of UK birds facing risk of global extinction, RSPBNews*. Available at: https://www.rspb.org.uk/about-the-rspb/about-us/media-centre/press-releases/kittiwakeredlist/ (accessed: 7 October 2021).

Haraway, D. (2015) ‘Chthulucene: Making kin’, *Environmental Humanities* 6, 159–165.

Haraway, D.J. (2016) *Staying with the Trouble: Making Kin in the Chthulucene*. Durham and London: Duke University Press.

Haraway, D.J. and Despret, V. (2020) *Phonocene, Bienal de Pensament Barcelona*. Available at: https://youtu.be/87HzPIEiF78 (accessed: 11 October 2020).

Ives, C.D. and Fischer, J. (2017) ‘The self-sabotage of conservation: Reply to Manfredo et al.’, *Conservation Biology* 31(6), 1483–1485. doi:10.1111/cobi.13025

Kanngieser, A. (2020) ‘Listening to Ecocide’, in *Sonic Acts Academy*. De Brakke Grond, Amsterdam. Available at: https://www.youtube.com/watch?v=dh1Cs4G7mxk

Kanngieser, A. and Todd, Z. (2020) ‘3. From environmental case study to environmental kin study’, *History and Theory* 59(3), 385–393. doi:10.1111/hith.12166

McCartney, A. (2014) ‘Soundwalking: Creating Moving Environmental Sound Narratives’, in S. Gopinath and J. Stanyek (eds), *The Oxford Handbook of Mobile Music Studies, Volume 2*, pp. 1–27. doi: 10.1093/oxfordhb/9780199913657.013.008

Ouzounian, G. (2021) ‘The Sonic rewilding of cities: Listening after Lockdown’, in (EHESS), T.M. & I. (ed.) *Crafting a Sonic Urbanism: Listening to Non-human Life*. Paris.

Paulson, S. (2019) ‘Making kin: An interview with Donna Haraway’, *Los Angeles Review of Books*. Available at: https://www.lareviewofbooks.org/article/making-kin-an-interview-with-donna-haraway/ (accessed: 11 October 2021).

Plumwood, V. (1993) *Feminism and the Mastery of Nature, Feminism and the Mastery of Nature*. London and New York: Routledge.

Plumwood, V. (1996) ‘Being prey’, *Terra Nova* 1(3), 32–44.

Radicchi, A. (2017) ‘A pocket guide to soundwalking’, in A. Besecke, A., J. Meier, R. Pätzold, and S. Thomaier (eds), *Perspectives on Urban Economics: A General Merchandise Store*. Berlin: Universitätsverlag der TU Berlin, pp. 70–73.

Randrup, T.B., Buijs, A., Konijnendijk, C.C., and Wild, T. (2020) ‘Moving beyond the nature-based solutions discourse: Introducing nature-based thinking’, *Urban Ecosystems* 23(4), 919–926. doi:10.1007/s11252-020-00964-w

Redfern, C.P.F. and Bevan, R.M. (2014) ‘A comparison of foraging behaviour in the North Sea by Black-legged Kittiwakes Rissa tridactyla from an inland and a maritime colony’, *Bird Study* 61(1), 17–28. doi:10.1080/00063657.2013.874977

Ruiz Arana, U. (2019) ‘The wild in silence’, *Interference Journal* (7 Sound & Environment: Sense of Place). Available at: http://www.interferencejournal.org/wp-content/uploads/pdf/Interference Journal - Issue 7 - The Wild in Silence.pdf

Rutz, C. *et al.* (2020) ‘COVID-19 lockdown allows researchers to quantify the effects of human activity on wildlife’, *Nature Ecology and Evolution* 4(9), 1156–1159. doi:10.1038/s41559-020-1237-z

Schafer, R.M. (1977) *The Soundscape: Our Sonic Environment and the Tuning of the World*. New York: Knopf.

Schultz, H. and van Etteger, R. (2017) ‘Walking’, in A. van den Brink, D. Bruns, H. Tobi and S. Bell (eds), *Research in Landscape Architecture: Methods and Methodology*. Oxon: Routledge.

Tong, H., Aletta, F., Mitchell, A., Oberman, T., and Kang, J. (2021) ‘Increases in noise complaints during the COVID-19 lockdown in Spring 2020: A case study in Greater London, UK’, *Science of the Total Environment* 785 (December 2019), p. 147213. doi:10.1016/j.scitotenv.2021.147213

Tozer, L. and Xie, L. (2020) *Mainstreaming Nature-Based Solutions: Climate Change*, NATURVATION Guide. Available at: https://naturvation.eu/system/files/mainstreaming_nbs_for_climate_change.pdf

Tyne Kittiwakes (2021) *Tyne Kittiwake Colonies, Tyne Kittiwakes*. Available at: https://www.tynekittiwakes.org.uk/tyne-kittiwake-colonies/ (accessed: 7 October 2021).

Van der Jagt, A., Toxopeus, H., Dorst, H., and Dévényl, R. (2020) *Mainstreaming Nature-Based Solutions: Economic Regeneration*, NATURVATION Guide. Available at: https://naturvation.eu/system/files/mainstreaming_nbs_for_economic_regeneration.pdf

Vincent, C. (2019) 'Donna Haraway: "Avec le terme chthulucène, je voulais que l'oreille entende le son des terrestres"', *Le Monde*. Available at: https://www.lemonde.fr/idees/article/2019/01/31/donna-haraway-la-pensee-chthulu_5417206_3232.html

Westerkamp, H. (2007) 'Soundwalking', in Carlisle, A. (ed.) *Autumn Leaves: Sound and the Environment in Artistic Practice*. Double Entendre.

Westerkamp, H. (2015) 'The Disruptive Nature of Listening, Keynote Address International Symposium on Electronic Art'. Vancouver, BC. Available at: http://www.sfu.ca/~westerka/writingspage/articles pages/disruptive.html (accessed: 19 October 2015).

Wolch, J. (1996) 'Zoöpolis', *Capitalism Nature Socialism* 7(2), 21–47. doi:10.1080/10455759609358677

Wright, K. (2014) 'An ethics of entanglement for the Anthropocene', *Scan, Journal of Media Arts Culture* 11(1). Available at: http://scan.net.au/scn/journal/vol11number1/Kate-Wright.html

2

WALKING WITH SOUND: RACE AND THE PROSTHETIC EAR

Louis Chude-Sokei

African American writer Ralph Ellison's classic 1955 essay, "Living with Music" matters much to me not just because of its famous insights about Black music and American culture in the year after racial desegregation was legally achieved by the historic Brown Vs the Board of Education case. But because it historicizes Black listening practices in ways that profoundly shape my own scholarship and writing as well as my work as a field recordist and sound maker who works primarily with spaces marked by race and historical trauma. The essay begins by establishing the stakes of race, sound, and listening in a quite personal way: "In those days it was either live with music or die with noise, and we chose rather desperately to live" (Ellison, 1972, p. 187). Due to this intimacy that blends the scholarly and critical register with the personal and autobiographical, the essay serves as a necessary introduction to my work since at the core of that work is the practice—well, habit—of soundwalking.

Soundwalking began informally for me, as a result of and response to my own peregrinations from country to country years before encountering the work of Ellison or composer, writer, and environmentalist R. Murray Schafer, to whom we owe the field of Acoustic Ecology and, along with composer and artist, Hildegard Westerkamp, the very notion of a soundwalk. As a habit, it also preceded my research into race, sound, and technology, the three elements which fatefully came together in the history of field recording. Field recording was important because I routed its history through the colonial practices of ethnography. Due to its occasional dependence on sound recording, ethnography was an early example of the technologization of cross-cultural encounters. It was also arguably the origins of our racialized modes of listening.

Ellison's essay is not about soundwalking or field recording, nor does it make overt reference to Brown Vs the Board of Education. But it signals that latter event by expressing what would matter for my work most: his insights about sound and

DOI: 10.4324/9781003193135-3

listening in a racially stratified and segregated nation. His was a nation where Black music, Black sound would have to bear the responsibility of crossing the cultural and social borders that Black people themselves would be forbidden to cross even after Brown Vs the Board of Education. That is why the politics of assimilation—and creolization—can be heard no clearer than in his aesthetically true yet perhaps politically optimistic statement that 'in the United States when traditions are juxtaposed they tend, regardless of what we do to prevent it, irresistibly to merge' (Ellison, 1972, p. 197).

That masterful essay, first published in the magazine *High Fidelity*, should by now be a foundational work of what we now call Sound Studies; if not, then certainly, Black Sound Studies. It marks a crucial shift from a by then established history of Black sound-making and the nascent canonization of jazz to a focus on communal audition in contested social spaces. But even more centrally the essay explicitly marks its territory as that which exists beyond or outside of music—sound.

For those unfamiliar with the piece, it features a relentless back and forth between Ellison the former musician and aspirant writer, his neighbors, and indeed his entire Black urban community:

> To our right, separated by a thin wall, was a small restaurant with a juke box the size of the Roxy. To our left, a night-employed swing enthusiast who took his lullaby music so loud that every morning promptly at nine Basie's brasses started blasting my typewriter off its stand.
>
> *(Ellison, 1972, p. 187)*

But it is his upstairs neighbor who catalyzes his complaints, transforming them into the foundational statement of competitive, auditory sociality that his essay is—a female vocalist whose material ranged from a "mastery of the *bel canto* style, German *lieder*, modern French art songs and a few American slave songs" (Ellison, 1972, p. 190). It's worth fully describing what Ellison hears because that's where audition is emphasized and where sound and listening become increasingly centered in shared public space:

> From morning to night she vocalized, regardless of the condition of her voice, the weather or my screaming nerves. There were times when her notes, sifting through her floor and my ceiling, bouncing down the walls and ricocheting off the building in the rear, whistling like tenpenny nails, buzzed like a saw, wheezed like the asthma of a Hercules, trumpeted like an enraged African elephant—and the squeaky pedal of her piano rested plumb center above my typing chair.
>
> *(Ellison, 1972, p. 190)*

Music is secondary here, its resonances transformed into sound and materialized. Overwhelmed, Ellison decides to fight "noise with noise" despite the "ethical problem" of being an artist complaining about the work of another (Ellison, 1972, p. 193). He doesn't depend on his skills as a trumpet player to fight back, especially since yelling and banging on the wall proved unhelpful. What works comes from his discovery of what Jamaican reggae and dub producers would call "the implements of sound" (Chude-Sokei, 2018).

The reference here to Jamaican sound culture is neither random nor accidental, given that it is that culture that would introduce me to soundwalking in advance of reading Ellison's *Shadow and Act* (1964) where I encountered 'Living with Sound.' In Ellison's case, those "implements" included the "booby trappings of audio equipment, wires, discs and tapes" and they allowed him to blast his upstairs neighbor with vinyl records (Ellison, 1972, p. 187). Perhaps this is where we can see traces of a masculinism that would equate Black maleness with a mastery of the technologies of sonic reproduction and Black femaleness as a primarily acoustic, as in supposedly organic or unmediated, phenomenon.

Perhaps. It is certainly where we hear Ellison charting that early/mid-twentieth century shift in Black listening practices and their attendant imaginings of racial community from the acoustic to the electric, the rural to the urban. This shift was a product of changing Black and American soundscapes, particularly as migrants left the agrarian, Jim Crow South and migrated into a world defined by less physical space and more electricity. The latter would reshape their music and the cultures around it. As he exclaims, "Between the hi-fi record and the ear, I learned, there was a new electronic world" (Ellison, 1972, p. 194).

Most important for field recording and soundwalking, which are for me twinned practices, is precisely Ellison's emphasis on technology as well as his deployment of sound and sound recording gear in outdoor/communal/public space. Though soundwalking need not depend on microphones or recording equipment, for me both practices emerged from a relationship between listening and technology, as I will later describe. They shaped a relationship between my ears and what Lawrence English called the prosthetic ear of the microphone, aided, and abetted by increasingly accessible recording equipment (English, 2015, p. 1). Black music is an index of this accessibility.

The sonic back and forth, played out in the cramped spaces of the apartment complex in which he and his sonic nemesis lived, describes a vision of competitive sociality that mirrors the "cutting contests" and sonic competitions among jazz musicians he documents in this and so many other famed essays on Black music. When I first read "Living with Sound" it was clear this wasn't just the work of one of America's greatest writers. Ellison was one of our great listeners. That his listening wandered and migrated, room to room, inside and outside, apartment to street, was familiar to me. The ambience, or more appropriately the "soundscape," of Ellison's community would return to me upon reading the work of R. Murray Schafer. After all, it was Schafer who proclaimed, "The blurring of the edges between music and environmental sounds may eventually prove to be the most striking feature of all twentieth-century music" (Schafer, 1994, p. 111).

The loudest island in the world

The language of Ellison's listening should be quoted at length to make clear how this essay could resonate across the Atlantic, beyond the world of Jazz or even Black America to reach an exiled listener in Jamaica. This listener—me—knew even at an early age that the Black Diaspora was as shaped by sound as it was by race.

Ellison writes,

> If, let us say, she were singing "*Depuis le Jour*" from *Louise*, I'd put on a tape of Bidu Sayao performing the same aria, and let the rafters ring. If it was some song by Mahler, I'd match her spitefully with Marian Anderson or Kathleen Ferrer; if she offended with something from *Der Rosenkavalier*, I'd attack her flank with Lotte Lehmann. If she brought me up from my desk with art songs by Ravel or Rachmaninoff, I'd defend myself with Maggie Teyte or Jennie Tourel. If she polished a spiritual to a meaningless artiness I'd play Bessie Smith to remind her of the earth out of which we came. Once in a while I'd forget completely that I was supposed to be a gentleman and blast her with Strauss' *Zarathustra*, Bartok's *Concerto for Orchestra*, Ellington's "Flaming Sword," the famous crescendo from *The Pines of Rome*, or Satchmo scatting, "I'll be Glad When You're Dead" (you rascal you!). Oh, I was living with music with a sweet vengeance.
>
> *(Ellison, 1972, pp. 195–196)*

That the turntable would be used as a primary instrument, competitive with, if not on par with an organic performance, would be meaningful for me, as would the suggestion that listening was neither passive nor innocent. As Ellison is honest enough to admit,

> For instead of soothing, music seemed to release the beast in me. Now when jarred from my writer's reveries by some especially enthusiastic flourish of our singer, I'd rush to my music system with blood in my eyes and burst a few decibels in her direction. If she defied me with a few more pounds of pressure against her diaphragm, then a war of decibels was declared.
>
> *(Ellison, 1972, p. 195)*

Though the essay specifies the songs and musical forms involved in this back and forth, what is most important in this exchange is neither genre nor the quality of recorded performance. The ability to select between and among songs and genres and performances is clearly important to this sonic combat; but most important is volume, and the ability to modulate sound in public space and deploy volume and pressure as power and discourse. All these elements became racialized in that they were tinged with the desire to blend—or *merge,* as Ellison would have it—as well as the enduring habit of fighting back.

Growing up in Jamaica, I came of age in a culture obsessed with sound, specifically bass, and volume. Ellison's "pounds of pressure" and "war of decibels" resonated powerfully. On this racially stratified Caribbean Island, music and sound blended openly with issues of violence, and masculinity, and public space was always contested, often violently. One could not help but walk with sound because silence was so rare as to seem impossible. Hearing the island referred to by its residents as the loudest island in the world, and its capital Kingston as the loudest city in the world, was a common statement of cultural pride and complaint whether it was true or not.

In my reading, then, "Living with Sound" merely described what was called a "sound clash" on the island I grew up in after arriving there from West Africa in the very early 1970s. That this culture would send migrants to the United States in that same period, who would inspire the culture of Hip Hop and its Dj/turntable battles would be no accident. Some would call this period the golden age of the island's most cherished cultural institutions, the Sound System. More specifically, the essay describes a *dub fi dub* session where selectors (those who selected and spun the vinyl records for a sound system) battled with each other in the island heat. Some of these battles could get violent as embarrassed rivals sabotaged equipment or enthusiastic gunmen fired shots of appreciation in the air. So, before being exposed to critics and theorists of sound and warfare, it was clear that sound was or could be violence and/or a form of resistance simply by walking through the streets of the island's major cities and on many beaches as well.

These mobile discotheques emerged on the eve of decolonization and were products of a predominantly Black working class who had little access to technology or control over colonial media. Sound systems may have begun as little more than glorified jukeboxes but, by the moment of independence from Britain, had evolved into massive and elaborate institutions with towering speakers, built and engineered mostly by the natives themselves. And in playing music chosen and then produced by the Black masses, these cherished cultural institutions could claim some degree of responsibility for the process of cultural decolonization that began before formal independence.

Sound system events were largely outdoors in Jamaica, and so they were inescapable. Tune for tune, back and forth at volumes so high the space between music and environment was blurred, especially given that audience responses were expected and incorporated into the soundscape along with a native obsession with echo and reverb. These effects would begin to define the music, particularly that subgenre of reggae that would become central to my listening, my writing, my research, and my recording—*dub*.

This obsession with echo and reverb was more than simply a fetishizing of sonic gimmickry or else it would not have become definitive of the music of that era. It clearly spoke to some cultural need. There was something there about the joys of exploring technology via sound but also about redefining and claiming space, and about self-expansion and duplication. Echo, for example, suggested so much about space and the manipulation of time and memory and sound. As a child on the island,

I remember my friends attempting to speak in echo like the great performers of that era who replicated analog echo effects in their vocal style, singers like Michael Rose, Echo Minott, the great Jacob Miller, and Eek-A-Mouse. And in response, producers like Scientist, Peter Chemist, King Jammy, and Lee "Scratch" Perry would suffuse the music in thickets of echo, dense like mangrove swamps. It is because of this shared obsession that I eventually named my overall field recording and sonic archiving project *Echolocution*, the dialect of echoes, the language of space.

At times it seemed as if the very environment was replaced by this extreme sonic activity, much to the consternation of those living on the peripheries of bass and those others who could afford to shield themselves with silence. It's worth pointing out the obvious fact that these institutions were not called *music systems* despite being ostensibly a delivery system for music, initially to people with less access to radios and stereos and who rightly acknowledged that island media was racist and colonial. It was the Black refusal to engage colonial radio—most notably the BBC and its island variants—that fed sound systems and the culture that quickly formed around them, which became sites of alternative social organization as well as supporting and fomenting resistant cultural movements, from the criminal rude boy sub-culture to the Rastafarians for whom the power of sound was a religious tenet.

They were also not called *reggae* systems. Sound systems were explicitly focused on sound, as was Ellison's great essay. They were where the "booby trappings of audio equipment" were fetishized, as signs of a modernity that could be claimed, rewired by native abilities, and shaped by what they assumed was ancestral wisdom. Technologies like microphones and speakers, and eventually mixers and effects. Sound men kept the gear pristine, at least until the dust of the dance covered the subs and tweeters where apprentice "sound boys" would fight each other for the right to keep those clean.

Outside of churches or perhaps on television or on record album covers, it was in sound systems where one first encountered these machines. They became signs of power, particularly for young men, and mastery of them a sign of masculinity. As a young listener in Jamaica and in Caribbean immigrant communities in America and England after leaving the island, the obsession with wires, knobs, levers, and machinery characteristic of sound system operators would follow me. As would their practices of listening. This obsession shaped my work because, as with Ellison, listening without the ability to respond and record seemed incomplete. A sound required a counter-sound, just as in the Black Diaspora a call required a response. That response was in this case a technological one and could be amplified and manipulated. Therefore, on the island and throughout its Black immigrant communities who spread the culture of sound systems globally, the obsession with technology was and is indistinguishable from the love of music.

Race and the soundscape

Years now after my encounter with Jamaica, R. Murray Schafer and dub-masters/sound system operators like the iconic King Tubby, Scientist, or Lee Scratch

Perry, I can reduce the insights of Ellison's essay into something epigrammatic or aphoristic. Music teaches us to listen, but it merely prepares us for sound. And sound prepares us for something else, which continues to guide my use of the prosthetic ear as I listen to and for relationships between space, sound, and race. R. Murray Schafer names this ineffable something else in the introduction to his famed treatise from 1977, *The Soundscape: The Tuning of the World* when he declares, "All research into sound must conclude with silence" (Schafer, 1994, p. 12).

Of course, silence is impossible. As any field recordist, experimental composer, or many soundwalkers would agree, it is merely a word used to describe the inaudible. It describes the myriad of sounds we shut out. Possibly the first thing one learns when one begins field recording is that most listening is silencing, a filtering out of information unconducive to our preferred or expected auditory foci. In Lawrence English's words, the prosthetic ear "fail[s] to extract signal from empirical noise" and so on playback, upon engaging the machine's listening, one is immersed in a soundscape shaped largely by our prejudices (English, 2015, p. 1). We are familiar with this process of bias when it comes to the visual register. We in fact have much language for the interplay of the visual and many types of bias or discrimination. However, the sonic equivalent of these forms of filtering can be trickier to acknowledge, hence the value and potential of field recording and soundwalking. They bring attention to what we do not, or at times refuse to, hear and they can make that refusal the object of consideration.

On the edges of a sound system—provided you didn't become interpellated by the aural regime of another—your hearing was exhausted by the high wattage and wide frequency range. Bass pressure deafened the ears, which worked to amplify physical responses to the sound. Only echo and random filter sweeps cut through the rumble and thud. Upon returning to volitional hearing, details became clearer. Exhausted and deafened, one's ears worked harder to make sense of the soundscape and to become more receptive to what had previously been filtered out of it. The moment the ability to note fine details returned, the pleasure was palpable, like sudden subjectivity.

Due to this cultural training, one couldn't help but become more sensitive to quotidian sounds and all the details that had been silenced due to bias or familiarity. In my case, I worked harder to listen to the marketplace and seashore, "ratbats" hurling past our ears in the night dark, crickets, peanut sellers, and, in Nigeria, endless village songs, and bottles clinking together gently but loud enough to signal the roadside presence of illegal brew. Or when I arrived in America, boomboxes from the windows of city busses or the deep bass rattle of souped-up lowriders in South Central LA. It was likely these dynamic contrasts that inspired me to spend hours on the edges of sound systems or African markets or public music events, or street corners, walking through these and other environments with half-closed eyes. These were my first soundwalks.

Again, soundwalking and field recording are twinned practices. Not just for me, but for many, especially with the advent of affordable, high-quality microphones and digital recorders. Not to mention smartphones and the range of sophisticated recording/editing applications that have enabled these practices to leave the rarefied provinces of anthropology, ethnography, sound art, and experimental music. As they should, especially since those practices have generated cultural spaces that can alienate listeners of color even though unofficial forms of soundwalking and rituals of non-musical listening are common to many cultures. There is much to be gained by these practices becoming mundane in the way applications like Instagram, TikTok, Facebook, and others have given rise to a global culture of visual *flânerie*. Admittedly, this culture is highly mediated and shot through with voyeurism and forms of spectatorship reminiscent of colonial encounters. But it is also true that these visual and technological cultures have been embraced and innovated by people of color across the planet, just as sound recording has.

The popularity and ubiquity of these new and highly accessible visual applications should remind us that the focus on listening that defines field recording and soundwalking can allow listeners to explore realms of affective and political power that the image has either lost or seen wither due to that ubiquity. That is why it's important to keep in mind that a politics of sound was already built into field recording due to its nineteenth-century origins in colonial, cross-cultural encounters. It is where machines—beginning with wax cylinders—attempted to collapse the geographic sprawl of empire and make it sensible, immediate, intimate, and affective for the white listener in ways the visual register could not. The prosthetic ear, or mechanized, non-cognitive listening attempted to render legible peoples and cultures organized by colonial power. It aurally reduced the sprawl of empire and its echoes into an apprehensible and containable whole with an intimacy that visual images could not replicate.

This political history was even more clear to me after research into the history of field recording, particularly as it would help construct a notion of "world music" in the wake of the Berlin School of Comparative Musicology in the nineteenth century. Ironically, much of the music played on sound systems was still called "world music" by Western consumers. But in Jamaica this music seemed eager to reject or critique colonial *worlding* by being explicitly about colonialism, migration, and racism. And those who made and consumed it developed and articulated associations between sounds and culture—for example, the bass as cultural and racial movement; drum rhythms as collective heartbeats; minor keys as "Ethiopian"; certain tones as "African"; and as always, echo as distance, memory, the loss of origins. Though steeped in racial essentialism and an often lethal masculinism, this was an alternative lexicon of listening generated by differential and politicized modes of listening.

Despite his own questionable racial attitudes, Schafer validated some of the above political concerns in *The Soundscape* when he establishes his goal of determining "in what significant ways individuals and societies of various historical eras listen differently" (Schafer, 1994, p. 151). As a product of social worlds that largely engaged in racial discourse through music, it was no great stretch to then reframe Schafer's goal with my own: to explore how different cultures and social groups listened, especially since differences in sound-making or listening were almost always attributed to race. And if sound could be coded in historical terms of colonialism, slavery, pride, resistance, and desire, could it also code or narrate distinct notions of the future? Remember, Schafer begins his influential work with a statement that Ellison would no doubt have found sympathetic to his own thinking: "the general acoustic environment of a society can be read as an indicator of social conditions which produce it and may tell us much about the trending and evolution of that society" (Schafer, 1994, p. 7).

Sound stains

With Ellison, Schafer, and my history with sound systems, the twinned practices of field recording and soundwalking became semi-formalized. This was after painfully acknowledging that music making had only prepared me for a more focused kind of cultural listening via recording technology. The goal became how to capture sound, but also silence because it too carried semantic, historical, and cultural freight, particularly the traces of all that had been silenced or filtered out. I wanted to know how best to record and read those worlds opened by hypersensitive microphones. Those worlds revealed themselves through what listening to silence reveals, which is primarily *space*.

We acknowledge that different cultures have distinct notions or systems of value that distinguish between noise and music. I was curious to learn if they had different conceptions of silence or absence. What did silence mean or trigger? How do different cultures, social groups, and communities conceive of their "general acoustic environment," or space itself? And how to record space in ways that capture the distinctiveness of an environment without the exoticism that still clings to field recording?

Space, after all, is what registers when principal sounds disappear from the field of perception or when the ambience around those sounds exerts enough pressure to impress themselves on the recording. As a field recordist and one who walks with sound, space is never empty, and emptiness is never immaterial—and I mean this more than just in terms of metaphor or associative cultural meanings. Sound, for example, travels differently in different temperatures due to humidity and the vagaries of atmosphere, and so silence—the inaudible—varies as much as does noise. Sound enters the body differently depending on location, reverberates differently (comparing subwoofers from sound systems in Jamaica with those in Brixton would teach me that). Microphones are sensitive to all of this, and operate differently from space to space, and so echo and reverb, or even equalization, are affected by location. I wanted to register and understand the cultural effects of these differences, not only in terms of how people in those places produce sound (music, language, echo) but how they relate to memory and space.

My research into ethnographic listening would only affirm my political goals and cultural interests because though we associate field recording with ethnic, indigenous, folk, and immigrant music, one of its most enduring products is in fact space. Not just the replication of the environs of the native via microphone placement or other on-site decisions in which one emphasized the foreign, the primitive. But as microphones became increasingly sensitive over the course of the twentieth century, and their directional capacities increased, the prosthetic ear couldn't help but capture the texture of life as represented by the space surrounding the primary sound object or subject. The room, the valley, the prickly heat, the storm became audible. These previous silences became even more noticeable in the epochal move from analog, tape-based recording to digital.

But it was no great leap from the recording of "ethnic" or "native" music and peoples from the late nineteenth century to the explosion of "exotica" in the wake of domestic hi-fi stereo technology in the 1950s, and to the popularization of field recordings in the commercial boom of "nature" sounds in the 1970s featuring oceans, environments, birds, and especially whales. Both represent forms of colonial or Eurocentric listening and/or auditory spectatorship. However, that latter work was also driven by ecological concerns, and it attempted to force consumers to revise a relationship to the environment that was rooted in the extractive dynamics of colonial power. One could at least find or hear in it some traces of a desire to revise or reject colonial dynamics.

In my thinking, such a rejection or revision is arguably built into field recording and soundwalking. Where the former is inextricable from its colonial history, the latter is well poised to critique and reshape the former by establishing different relationships to sound, space, race, and memory. The tension between the two practices partly explains my focus on spaces that have been impacted if not defined by historical trauma. Such spaces tend to be overdetermined by their histories—as field recordist Chris Watson has said, "events haunt spaces" (Watson, 2017). Another way of saying this is *history contaminates place*, and there may be no greater or more powerful way of experiencing hauntings, place, and history than by receptive postures of listening.

Watson's insight also reminds us that listeners are already contaminated by the histories, associations, and meanings of a particular location, and so the spaces I choose are not assumed to be neutral. In encouraging history to bleed into present listening—allowing space to speak, as it were—soundwalking enables a critique of the colonial postures of field recording where an environment and those who occupied it were often mere agents of the listener's personal and cultural fantasies.

It is well known that such postures were present in Schafer's practice and theory, which was riddled with a bucolic romanticism and an anthropocentrism that centered the perspective of the kind of listener and cultural subject he was. Though I do not believe that listening can ever truly be non-anthropocentric, and a black male listener can also be riddled with perspectival and social privileges, the kind of listening my work prizes is decidedly non-deterministic, migrant. Perhaps because my work as a writer and scholar depends so much on interpretive and ideological frameworks, I fetishize a listening that challenges even my strongest political inclinations. Such inclinations, after all, often emerge from modes of self-justification that we mask with ideology. Hence also my resistance to manifesto or formalized methodology. My practice uses field recording and soundwalking with and against each other to explore what Ralph Ellison discovers in the contested public space of involuntary listening described in "Living with Music": "I was forced to listen, and in listening I soon became involved to the point of identification" (Ellison, 1972, 193).

In other words, sound does not separate. It does not absolve. Sound stains. It forces an identification with the other and other histories despite the differential social, economic, racial, or cultural locations of the listener. Sound blurs self and other, the past and the present, and can be used to make history more palpable and affective. This is because listening implicates, which is the primary thing I seek to make audible. It renders us complicit in the joys and crimes of history via listening to spaces scarred by those joys and crimes and the memories of life in between. As such, history becomes inescapably intimate and shared. Indeed, simply inescapable. This is likely why Ellison is able to declare in "Living with Sound" that "Identification, after all, involves feelings of guilt or responsibility" (Ellison, 1972, 189).

References

Chude-Sokei, L. (2018) "Dr. Satan's Echo Chamber: Reggae, Technology and the Diaspora Process," *Popular Inquiry: Journal of Kitsch, Camp and Mass Culture*. Available at https://www.popularinquiry.com/blog/2018/9/12/louis-chude-sokei-dr-satans-echo-chamber-reggae-technology-and-the-diaspora-process (accessed: April 16, 2022).

Ellison, R. (1964/1972) *Shadow and Act*. New York: Vintage Books.

English, L. (2015) "Relational Listening: The Politics of Perception," *Contemporary Music Review*, 36(3). Available at: https://www.tandfonline.com/doi/full/10.1080/07494467.2017.1395141 (accessed: April 16, 2022).

Schafer, R.M. (1994) *Our Sonic Environment and The Soundscape: The Tuning of the World*. Vermont: Destiny Books.

Watson, C. (2017) "Chris Watson on the Spirit of Places," *Sonic Field*. Available at: https://sonicfield.org/chris-watson-spirit-places/ (accessed: April 16, 2022).

3

WALKING AND NARRATING SOUNDBORDERSCAPES: AN EXPERIENCE IN HONG KONG

Elena Biserna

I would like to start by situating myself: I am a White cis able-bodied woman with a European passport. This gives me a quite privileged position in terms of freedom of movement and, despite the fact that I have lived on different boundaries, I have never experienced real border restrictions. I was born in Italy, one of the countries that are more involved in the 'management' of migration fluxes to the European Union. I did my PhD in the small town of Gorizia. After the Second World War, a wall was built on its Transalpina Square to separate Gorizia and Nova Gorica, Italy and Yugoslavia, Western and Eastern Europe. This wall was demolished well before I arrived there, in 2004 (Popham, 2004), and in 2008, when Slovenia entered the Schengen space, free movement was allowed within the square and between the two countries. Before moving to Marseille, I spent several months frequently travelling between Italy and South France by train. Each time, I watched dozens of people getting off the train in Ventimiglia; we then stopped before Menton, sometimes for an hour or so, for border controls. My documents have never been checked. Marseille is in the middle of the Mediterranean and was called the 'Gateway to North Africa'. In the ancient world, the Mediterranean often represented a connective space of economic, social, cultural and political exchange; today, European politics transformed it into a liquid but inaccessible entity that separates North and South; Europe, Africa and the Middle East: a border defining so-called 'Fortress Europe'.

The rhetoric of globalisation presented us with the image of a world without borders, where nation states had lost their importance and the free flow of goods and people was taken for granted. Nevertheless, the always-new 'migrant crises' that keep repeating themselves have clearly shown, even to people like me – who have experienced borders always from the 'right side' and with good documents – that this is a fiction. Today is 15 November 2021, and, while I write, thousands of asylum seekers and migrants are stuck in the woods in the middle of the

DOI: 10.4324/9781003193135-4

Poland–Belarus border crisis. More in general, as the philosopher Paul B. Preciado writes, since at least the fall of the two towers, we have witnessed a 'return … to the idea of the physical border as a condition for restoring national integrity and political sovereignty' and Europe and the United States have built themselves as spaces 'closed' to migrants (Preciado, 2020). At the same time, cultural identities are both reinforced and diluted through global processes and resistances that manifest themselves in very different, ambiguous and conflictual ways, often becoming a site of struggle.

Although I have always embraced the 'no borders, no nations' slogan, I first took the time to plunge into border studies and reflect on the new forms and configurations of borders some years ago, thanks of the invitation of the French collective *antiAtlas des frontières*.[1] I realised that, in our globalised world, paradoxically, borders proliferate and multiply while becoming more fluid and infiltrating new spaces and new systems, taking on new forms and acquiring an unprecedented significance in every field of our life.

Borderscapes and SoundBorderscapes

We used to imagine borders as dividing lines separating different territorial entities considered as homogeneous within them: the nation states. This is the image conveyed by maps such as the ones that we study at school, where each country is identified by a different colour and surrounded by a line. Indeed, this binary way of imagining territories corresponds to the modern logic of nation states but also to the Cartesian conception of space and to the cartographic device. This conception of space is neither neutral nor transparent. Cartography is a form of representation that was created at a specific time (during modernity and colonialism) and in Europe. As such, it expresses and produces a specific view of/on the world (centred on Western Europe and functional to the colonialist project). As Sandro Mezzadra and Brett Neilson recall, 'it is important to remember that mapping was a key tool of colonial domination' (2013, p. 15). According to them, the linear image of the border established as part of this process assumes a particular significance today, materialising in the many walls that we see around the world and leading us to imagine borders first and foremost as a device of exclusion, focusing on the dichotomy between interior and exterior.

Nevertheless, first of all, different borders (geopolitical, linguistic, symbolic, cultural, economic and now sanitary) coexist; sometimes they overlap, but sometimes they don't. The logic of nation-states made all these borders coincide with the boundaries of their territories, but today this 'alignment' has the tendency to break apart. The philosopher Etienne Balibar points out that 'some borders are no longer located on borders'[2] (Balibar, 1994, p. 7), becoming increasingly mobile and pervasive. They materialise in places that are sometimes very far from the lines of separation between countries: airports, biometric recognition systems, refugee camps, courts, surveillance systems integrated into our mobile devices … In short, in any context marked by practices of control and selection.

Mezzadra and Neilson stress that borders do not have the function of blocking global flows, but that they have become central elements of their articulation. They divide but also connect: they filter and organise the passage of people, money, goods and objects. They allow certain forms of circulation and mobility (cheap labour, financial exchanges, mobility of artists and researchers, etc.) and prevent others (mainly economic or political migration) and this articulation is determined in large part by the neoliberal logic. These transformations are in fact closely linked to a project of freedom of movement which follows an economic agenda and creates different forms of exclusion according to its objectives (Mezzadra and Neilson, 2013, p. 7).

A notion that I found particularly generative in order to account for these multiple dynamics is that of 'borderscape' that, in a conversation with Holger Pötzsch, the anthropologist Chiara Brambilla defines as such:

> Differently from boundaries of nation-states that, according to the modern territorialist geopolitical imaginary, are invented as lines on the flat and two-dimensional surface of the map, borderscapes are multidimensional and mobile constructions that are inhabited by Michel de Certeau's 'spatial practices'. [...] In the borderscape, representations and practices produce a multiplicity of interrelations that call into question modern state-centric geographies and politics of identity. The ability of the borderscape to bring together experiences and representations helps to develop a performative approach to borders. The borderscape concept allows us to move beyond the often-criticized gap between practices and representations and enables us to abandon the essentialization of borders as divisive lines on modern political maps, moving instead towards an understanding of contemporary borders as continually performed and (re)composed by sets of contingent performances revealing their dynamic character.
>
> *(Pötzsch and Brambilla, 2015, n.p.)*

Brambilla describes a border that is performatively built through our spatial practices and representations and generated by the confrontation of (violent) bordering institutions and processes of border-crossings. In this way, she also underlines the agency of practices of resistance and border transgression, rather than insisting on metaphors and conceptualisations that suggest both the exclusion of 'others' and the 'homogeneity' of what is included by borders.

Starting from this notion and taking into consideration the role images, imaginaries and narratives have in shaping both the meanings and effects of borders on people's lives (Strüver, 2005), I focused on representations and interventions based on sound. By using the term 'SoundBorderscapes', I wanted to propose that listening could be a tool to question the traditional conception of the border as a line of separation as well as to complexify the identity and supposed homogeneity of what is included by this line. I analysed several projects by artists such as Justin Bennett, Ultra-red and Lawrence Abu-Hamdan recording borders, composing border struggles or critically analysing the politics of voice and language in specific territories, realising that these 'Soundborderscapes'

> deconstruct the dualism between the inside and the outside to offer an 'image' of the border as a membrane constantly penetrated by the other's crossing and, at the same time, articulated, constituted and transgressed by a multitude of practices, events, discourses and relations. In this way, they not only amplify and 'make audible' the processes that take place at the border – or the border itself as a process – but also propose listening as a useful instrument for a critical epistemology of the border.
>
> *(Biserna, 2017, n.p.)*

Since Cartesian conceptions of space crystallised in maps continue to insist on the inside/outside dichotomy and influence our way of thinking and acting, I wanted to suggest that listening could be generative to grasp borders as 'real-time mobile flows' (Popescu, 2013), questioning the oppositions between belonging and otherness, inclusion and exclusion through the vibrant, dynamic, contingent, plural and relational matter of sound. As Salomé Voegelin also observes:

> the sonic verticality of a continuous simultaneity makes apparent that the discrete is an illusion and that partitioning is just the admission of an insurmountable inter-existence. Against this admission, a geography of sound produces a different political imagination that focuses not on dividing but on interbeing and the sharing of the voluminous expanse of the world. [...] The sonic articulates an imagination of indivisibility and questions the logic of separation, producing a different political possibility of a cosmopolitan interbeing that acknowledges our insurmountable simultaneity.
>
> *(Voegelin, 2019, p. 92)*

Hong Kong borderscapes

With these reflections in my mind, I had the opportunity to go to the border for a collective practice-based project. In April 2019 I was invited by soundpocket – an art organisation focusing on sound in Hong Kong – to work with a group of local artists for a mentorship programme. When I received this invitation, I immediately thought about the specificity of the geopolitical situation in Hong Kong. Therefore, I proposed to work collectively on its identity politics and to do it by exploring its borders. By taking into account the 'symbolic power' of borders in organising cultural difference and their 'interiorisation by individuals' as 'an essential reference point for their sense of collectivity and community and therefore ... of their identity' (Balibar, 1994), I wanted to start a critical inquiry on the different forms of border both inside and at the margins of its territory to address the so-called 'Hong Kong identity crisis'. The more I talked with soundpocket and researched, the more it was clear that the border in Hong Kong manifests itself in a plurality of places and ways: on the material walls and fences, in the different land crossings, in language politics, in the multiple control and selection apparatuses in urban space, in passports and documents and, first of all, in collective and individual representations, imaginaries, memories and identifications.

The literature scholar Ackbar Abbas speaks of 'the cultural self-invention of the Hong Kong subject in a cultural space' that he calls 'space of disappearance' (Abbas, 1997, p. 1). According to him, the national identity and cultural specificity of Hong Kong have been produced through a process of differentiation with China at the moment when its boundaries were fading. A global, colonial city, a world financial centre, a port, a target of migration from China and full of expats, Hong Kong has always been a place of transit and has never needed a precise local identity until the risk of homologation with China became urgent. This happened in 1997 with the handover (when the 157 years of British rule on the colony were ended with the transfer of sovereignty to the People's Republic of China): a fracture in Hong Kong history, but also on many individual lives.

Starting from readings and regular exchanges with the soundpocket team, I decided to propose a programme based on on-site visits and meetings, on experience and praxis, using listening, walking, sound collecting and storytelling as methodologies to question material and immaterial borders, to explore their polysemy – the ways they are perceived by and affect different social groups (Balibar, 1994) – as well as the role they play in the production of subjectivities and identities. Listening and sound collecting were imagined as tools for experiencing the multiple, porous and dynamic nature of borders. Storytelling was meant as a way to reveal the singularity and plurality of our identifications, starting from heterogeneous perceptions and experiences of borders and the way they assume different meanings and have different repercussions even for people who live in the same territory; collecting stories as a way to grasp the effects of borders on the lives, freedom of movement and very existence of people with different histories, statuses or intersecting factors, starting from their passport.[3] Walking was proposed as a practice of 'ambulatory knowing – or knowledgeable ambulating', as suggested by Tim Ingold (Ingold, 2000, p. 230) or as a methodology for 'knowing the world through the body and the body through the world', as Rebecca Solnit says (Solnit, 2001, p. 29).

Over the last century, walking has been the subject of a vast interdisciplinary literature: from the social sciences to anthropology, from urban studies to cultural studies, from philosophy to art theory and history. In spite of their variety, these studies converge on at least one point: the impossibility of reducing walking to a purely functional activity allowing us to go from point A to point B. On the contrary, walking is interpreted as a privileged practice for questioning our relationships to environments and territories: a dialectical interaction with the very mobility of sites and their physical and immaterial transformations. Thus, the micro-practice of walking generates a constellation of relational possibilities. It allows us to observe space (Benjamin, 1982), but also to rewrite and redefine its structure (Barthes, 1985; de Certeau, 1980; Augoyard, 1979; Paquot, 2008). It establishes an embodied and situated exchange with the environment (Sansot, 2000; Le Breton, 2000; Thibaud, 2008; Thomas, 2010), while providing an opportunity to participate in the public sphere and to encounter otherness (Jacobs, 1961; Solnit, 2001; La Cecla, 1988). Therefore, through walking and listening, I wanted to propose a way to experience and practise space through immersion and contact, providing a form of physical, subjective and situated spatial knowledge. In my view, this could help us to problematise the disembodied, universal, objective and objectifying perspective of the modern cartographic gaze that represents space as a homogeneous, measurable and fixed entity according to a two-dimensional Cartesian system. That is, to the conception of space on which traditional notions of borders are based. By combining these practices, our challenge was to rethink the image of the border in Hong Kong as a line of separation between two completely heterogeneous entities as well as to try to grasp the complexity and ongoing changes of these entities themselves. Taking inspiration from Ultra-red's *Militant Sound Investigation* (Ultra-red, 2014), I decided to introduce a question to guide our exploration: what is the sound of borders in Hong Kong?

The workshops

On the first day we all met in Hong Kong, we started to discuss together the notion of border – what is a border? How do we feel and experience it? How does it sound? I also invited all the participants to ask the same questions to their relatives and friends, to look for other personal, plural and possibly divergent stories. We listened to some artistic projects dealing with geopolitical borders through sound and listening, among which was Samson Young's *Liquid Borders* (2012) – a project comprising graphic scores, annotated cartography and compositions based on field recordings made with contact microphones and hydrophones on the restricted zones along the border between Hong Kong and Mainland China: the vibration of the fences and walls, the water of the Shenzhen River …

On the second day, we went to the same border to walk along it, listen to it, sense it, play it. Most of the border areas are inaccessible but, with the help of soundpocket, we found a route along the wall that we could safely follow for some kilometres. The group was heterogeneous and composed of artists with very different backgrounds. So, first of all, I introduced the notion and history of soundwalking. However, compared with the most widespread methodologies and canonised formats for soundwalking (see McCartney, 2010; Paquette and McCartney, 2012; Drever, 2020), I didn't impose a specific rhythm to our walk, letting each find their own pace and take their time. Some of the participants decided to take field recordings, lingering on some spots or moving away from the group from time to time; others decided to record their experiences in different ways, especially writing and taking notes. I also didn't put any emphasis on silent listening and rather invited the participants to 'play' the environment and engage with the site.[4] Considering *milieus* as fields of dynamic, temporary and processual relationships – as multiplicities constantly generated in relationship with our changing positions and behaviours – I am particularly interested in site-specific sound-making. As Barry Truax already highlighted in the *Handbook for Acoustic Ecology*, sound-making can expand our perception and engagement with sites generating a process of re-appropriation and interaction through action. For Truax, it was a matter of 'explor[ing] sounds that are related to the environment, and, on the other hand, [of] becom[ing] aware of one's own sounds (voice, footsteps, etc.) in the environmental context' (Truax, 1978). For me, it was more a way to feel our presence in such a contested place and activating its matter – the fences, the gates, the walls dividing Hong Kong from Mainland China – to generate vibrations with our own bodies and listen to the result of our action on its and our own physicality (Figures 3.1a–c).

FIGURES 3.1 a–c *Hong Kong SoundBorderscapes*, organised by soundpocket, Hong Kong, April 2019

After almost an hour of quiet and scattered listening, a subtle action on the fences generated a sound that we couldn't hear: an alarm at the police station triggered by the sensors on the walls. The police arrived in the middle of our soundwalk and interrupted it with their sirens. We were struck for a while and, when we finally could leave, I think our listening attention was more focused on specific sounds – dogs barking, airplanes – and we had a different feeling of the border.

This interruption helped us to deal with and speak about borders and their effects in more direct ways. During the long discussion that followed, when we reached the first village, a lot of reflections on our listening experience emerged, linking it more directly with the notions of welcome and unwelcome, authority and power, control and surveillance, alert and fear, safeness and unsafeness, familiarity and unfamiliarity. Of course, my position and orientation were very different from those of my comrades: those of a stranger that walks along a wall experienced mainly through readings; a wall that I had crossed just once, the day I arrived in Hong Kong, to get my working visa after waiting for a few hours to enter Mainland China through the Shenzhen Border Crossing. For most of the participants, this wall had completely different meanings and associations: memories and stories that marked their life and sense of identity. Some of them had never crossed it; others were born in other countries such as Australia – their parents had travelled in order to give them another passport that would have allowed them to resettle in the case that China were to have taken over; for almost everybody, this wall marked a safe space of freedom against the threat of a life under surveillance.

These reflections continued the next day. We had planned a soundwalk in the city centre to explore immaterial borders in urban space but the rain was so strong that we were obliged to spend the whole afternoon in a café close to the place where the Umbrella revolution had taken place in 2014. We took the occasion to discuss our personal experiences of border crossing using our narratives to deal with wider issues connected to inclusion and exclusion as well as fear and ambiguity. This forced stop was very generative and helped us approach a more critical vision and to deconstruct some of our preconceptions.

We started to discuss our languages, what we considered our mother tongue, and how this was important for our own identity. As I grew up in a little countryside village in Emilia Romagna, Italian is my mother tongue, but I also understand and speak – as many other people of my age and social class – our dialect. Italian as a national language is a quite recent phenomenon; it was universalised with the diffusion of television and dialects are still frequently spoken, especially in rural areas. Then I had to learn English – as most people of my education level given the hegemony of this language on a global scale – and I learned French when I moved to Marseille. Translation, for me, is thus part of my identity as much as my mother tongue. My personal experience led us to discuss language politics in Hong Kong: the fact that all the participants recognised Cantonese as their language and a relevant part of their identity, while Mainland China was trying to diffuse Mandarin, homologating Hong Kong through its official national language. But we also talked about some of the participants' relatives, who, when they arrived in Hong Kong, spoke other languages or different varieties of Chinese, and learned Cantonese as a way of 'belonging' to their new place of residence. We started to look at the way language organises difference and how nation states build themselves in and through languages, imposing unity on a (pre-existing) multilingualism.

Thanks to the proposal of one of the participants, our conversation switched to the sounds that define Hong Kong identity. What could be a 'sonic postcard' from this territory? We started to make a list of what soundscape studies name 'soundmarks'; that is, according to the definition of Barry Truax, 'a community sound which is unique, or possesses qualities which make it specially regarded or noticed by the people in that community'. Soundmarks, as Truax writes, 'are of cultural and historical significance and merit preservation and protection' (1978). By looking and discussing further, we realised that many of the sounds first mentioned as 'specific' to Hong Kong were actually linked to the colonial period, such as horse races, or to cultural practices that go across borders or come from different parts of Asia, taking different shapes as they travel.[5] This highlighted how soundscapes, in the same way as cultures, are never 'unique' or 'specific' to a community, but rather are marked by transfusions, cross-pollinations, *métissages*.

Thinking about languages and the sound of the city helped us to reflect on our ways of constructing our identity negotiating different kinds of borders. In other words, discussing the internalisation of borders: how we all somehow incorporate the vision and ideology conveyed by nation states by identifying with the figure of the national citizen, how borders enter the construction of our own subjectivity; to, I hope, complexify our vision and start to understand identity as an ongoing production and relational process, always embedded with difference and hybridisation, made through displacements and not only linked to our origin. In other words, identity as 'becoming' rather than 'being' (Hall, 1994). Then we went on talking about the relationship between borders and authority, the sense of being under control and surveillance, starting from personal experiences and feelings of fear and unease. This discussion was pivotal in order to reflect together on many different issues and how they are related to our listening experience.

Over the following days, we tried to find a distance from our own personal perspective to listen to how 'others' experience the borders in Hong Kong. We had another soundwalk in a racialised multi-cultural borough. This gave us a chance to meet people and start small informal discussions but also to listen to the heterogeneity of languages, accents and voices in Hong Kong and reflect on their status. Then we met activist Innocent Mutanga – a former asylum seeker from Zimbabwe, anthropologist and co-founder of the Africa Center. Talking with him allowed us to acknowledge the stories of people who are displaced in Hong Kong without the right to stay or the right to leave, and who are subject to multiple forms of violence and discrimination. Starting from his own experience as a Black man in Hong Kong, we had the chance to reflect on how less privileged communities and social groups – migrant domestic workers, racialised or undocumented people and their children, for example – are affected by borders in Hong Kong in relationships with law, civil rights and freedom of movement, but also with everyday racism and preconceptions.

On the last day each participant presented a recording, replying to the question 'What is the sound of the border in Hong Kong?' Starting from a collective listening session, we wrote key terms and notions and discussed associations, recurrences and contradictions, without trying to find a synthesis but drawing the lines of a common (auditory) vocabulary for the next part of our programme (Figure 3.2).

FIGURE 3.2 Hong Kong SoundBorderscapes, organised by soundpocket, Hong Kong, April 2019

A research trip to Hanoi

The next step was a research trip with the group. I had proposed to go to China to continue our inquiry on the other side of the border, but I finally had to renounce that idea, and we went to Hanoi, Vietnam instead. In order to pursue the work started in Hong Kong in such a different context, I proposed to focus more in depth on personal and collective identities and their complexity (Vietnam has 54 recognised ethnic and linguistic minorities) and to do so from a particular perspective: that of a stranger who arrives in a place they don't know – the same perspective I had when I arrived in Hong Kong. I wanted to make listening *derives*,[6] to focus on welcoming and unwelcoming sounds and to translate itineraries into experiential maps. Since I am convinced that listening is a powerful tool for opening up to others, a form of permeability, a practice of inclusion and hospitality, I had prepared a series of exercises for listening to others or for collective listening inspired by the scores of Pauline Oliveros and other artists but also by practices I have learned in bio-political meetings. And, starting from a set of questions that emerged after the first workshop, we had imagined meeting different people and groups to interview them. We began by discussing essentialist and non-essentialist visions of identity, making listening exercises, addressing methodologies for the interviews (Figure 3.3).

FIGURE 3.3 Hong Kong SoundBorderscapes. Research trip to Hanoi. Score activation

But everything changed. We arrived in Hanoi on Monday, 10 June, the day after the first demonstration against the extradition bill in Hong Kong. Two million people (one third of the Hong Kong population) were on the street. Already, on the second day, during the first break, we were stuck to our phones as a second demonstration was taking place and police were replying with brutal violence. On the third day, I decided to talk to the group: it was clear to me that we couldn't continue as if nothing was happening; we were physically in Hanoi but mentally in Hong Kong and the only way to keep working together was to make this experience of connection and disconnection the centre of our work: to transform our time in Hanoi into a space for remote listening, transmission and amplification of what was happening in Hong Kong. After discussing together, we decided to restart from this forced distance. Each participant found a way to translate this experience in a different form: recordings made in Hong Kong, discussions with relatives or friends, images of the demonstrations mixed with experiences in Hanoi in sound pieces, performances and video works. The presentation of the participants' works during an informal gathering at Hanoi Heritage Space turned into an occasion to talk about the situation in Hong Kong and make people aware of what was happening. For my part, I went for a walk along the Hoan Kiem Lake, broadcasting a field recording of the demonstration in Hong Kong shared by the artist Fiona Lee to infiltrate that sound in Hanoi city centre, where the right to protest doesn't exist. The participants also organised a spontaneous 'Hong Kong Story' sit-in and we spent an evening in the streets to share Hong Kong's voice in Hanoi and tell passers-by what was going on.

Situated narratives, 'situated listening'

I came to Hong Kong with the somewhat naive intention of listening to and walking on its borders to collectively challenge a representation that creates a vision of the world based on the dichotomy inside/outside; which divides us into citizens and non-citizens; which fixes and homogenises our identity according to a predefined model forged by the ideology of nation states that simplifies the complexity of the people living in a territory. I had 'good intentions' (Azimi, 2011) – that is, I wanted to engage with identity politics in Hong Kong. But what was my possibility and even legitimacy to do so as a Western cultural worker working with local participants? One and a half months later, the protests in Hong Kong erupted in order to defend one of the borders between Hong Kong and Mainland China: the independence of the Hong Kong legal system. Five months later, by November 2019, when I was safe in my apartment in Marseille, the struggle against the extradition bill had become a fight for freedom and against a police state. Borders also manifested themselves directly on the demonstrators' bodies, on the bodies of the Polytechnic University students, and resounded in teargas explosions, shots and sirens – in a soundscape of war that I could experience only in a mediated form, through videos, recordings and narratives.

This experience was really contradictory to me. From the other side of the world, I kept listening to voices from Hong Kong as a form of care, alliance and solidarity, but I also felt helpless and naive. Months later, my discussions with soundpocket took the form of a radio piece involving two people of the team, Mandy Chan and Vanessa Lai, along with Fiona Lee: an artist who works between performance and installation and who, since the beginning of the movement, has shared her field recordings of the demonstrations online. Together, we listened to these soundscapes and discussed the struggles and their transformations in a context of increasing control, especially after the national security law was approved in June 2020, strongly limiting hongkongers' civil and political freedoms. We focused, in particular, on the vocal tactics introduced by protesters: the role of songs and the 'Sing with You' flash mobs in propagating the struggles in everyday life, escaping control and the ban of political gatherings.

During this conversation, Mandy and Vanessa also underlined how, for them, it is important to keep recording and spreading these sounds. According to them, recording allows us to document and therefore to control the police's actions, while also contributing to catalysing energies and to raise awareness around the world. That's why they are committed to keep collecting and sharing sound recordings of the social movements in Hong Kong, despite the reinforcement of control and repression measures (Biserna and Rieussec, 2021). To me, these recordings – although not taken on official borders lines – are exactly SoundBorderscapes of struggle. They direct the microphone at the conflicts taking place around the border: 'those struggles that take shape around the … unstable line between the "inside" and the "outside", between inclusion and exclusion', as Mezzadra and Neilson put it (2013, p. 13).

The extradition law and subsequent social movements in Hong Kong made this line, once again, particularly unstable and yet visible and audible. Before coming to Hong Kong, I had imagined that walking and listening could help us access a contextual, relational and complex spatial knowledge of a conflictual borderscape, but suddenly the situation itself became even more complex and conflictual than we had imagined. The Hong Kong borderscape turned into a dissensual, agonistic *milieu* whose stakes were not possible to foresee. On the other side, the constant interruption of our walks and plans – the police's intervention, the weather and finally the protests – also obliged us to experience how soundwalking needs to be always open to the contingency of situations and their sudden transformations – an improvisation *with* sites rather than an attempt to compose them (McCartney, 2005; 2010). Moreover, these sometimes radically shifting situations and the discursive site of our soundwalks – the border(s) – drove us to account for the situatedness and subjectivity of soundwalking and the impossibility of a neutral or universal perspective: how our experiences were always necessarily different and strictly linked to our physical, symbolic, cultural and political position.

In this regard, while reflecting retrospectively, narratives and storytelling perhaps ended up being more productive in different ways: while in Hong Kong, listening to stories of displacement, migrancy and exile provided a strong tool to be able to grasp different border experiences and therefore asymmetrical processes of subjectivation, ways of belonging to, or feeling excluded from, a territory. This generated a multi-voiced account of the (heterogeneous and polysemous) experience of borders for people coexisting in the same territory with different statuses. While in Hanoi, the contingent situation made priorities change and the most urgent need for the group was to listen to the conflicts in Hong Kong to spread the word, explain what was going on, tell this story. This materialised in the 'Hong-Kong Story' sit-in, but also in the participants' works. Both these necessities – listening to other stories and being able to tell our own story – correspond to the primary need of any coexistence: to listen and to be heard (Bickford, 1996; Couldry, 2010).

During the final presentation in Hanoi, one of the participants said: 'During this trip I have learned to care about others and to listen more to others.' This intuitive association of care and listening is far from naive, and expresses an engagement, an opening, a sensibility. As Gemma Corradi Fiumara writes: 'the mechanism of saying without listening has transformed, multiplied and extended itself into generalised forms of global domination' (1985, p. 12). Making other narratives matters, creating spaces for them, listening to them, even (especially) when they expose our difference, might be a way of dealing with asymmetries and inequalities and also with the 'unstable lines' between the inside and outside, building solidarities and alliances or at least acknowledging our 'situated' positions (Haraway, 1988) and the other's 'right to opacity' (Glissant, 1997); that is, to be aware of our symbolic position in a complex network of relationships of domination and knowledge formation while, at the same time, recognising the unknowability and unintelligibility of the 'other', their right of not being reduced, essentialised or normalised according to preconceived epistemologies, visions and representations. This might at least generate a form of 'situated listening' (Biserna and Raimondo, 2022) as a vector of complexity and plurality to challenge hierarchies, normative discourses and systems of domination: a radical listening accepting ambiguity and singularity, the agency of the unheard and the imperceptible, the displacement of the I and of the paradigms through which we know the world. A listening looking for commons rather than differences. As Glissant writes: 'Thought of self and thought of other here become obsolete in their duality. Every Other is a citizen and no longer a barbarian. What is here is open as much as this there. … This-here is a wave and it waves no boundaries' (Glissant, 1997, p. 190).[7]

Notes

1 The *antiAtlas of borders* is a group of researchers, artists and experts focusing on the mutations of borders and spaces in contemporary societies. See: https://www.antiatlas.net/antiatlas-of-borders/
2 All translations of texts in languages other than English are mine.
3 This could be linked to development of 'biographical research methods' on the daily life of people in social studies considered, following O'Neill and Ismail, as a 'useful interpretative approach to identity' (2019, p. 34).
4 Elsewhere, I have written more on the role of sound-making in the early World Soundscape Projects soundwalks and map-scores as well as in contemporary performances and interventions (Biserna, 2021).
5 For a critical take on soundmarks from the colonial past, see Smolicki (2021).
6 The *dérive* is a practice of urban wandering introduced and theorised by Guy Debord and other members of the Situationist International. In the first number of the *Internationale Situationniste*, it is described as 'a mode of experimental behaviour linked to the conditions of urban society: a technique of rapid passage through varied ambiances' (IS, 1958). Within the IS, drifting became the main method for psychogeography, namely 'the study of the precise laws and specific effects of the geographic environment, consciously organised or not, on the emotions and behaviour of individuals' (Debord, 1955, n.p.). Moreover, it was also conceived as a revolutionary tool: a way to subvert the functionalism, consumerism and alienation imposed on everyday life and urban planning through a collective practice of disorientation, an experimental way of inhabiting places, and a different use of space and time. See also Sadler (1999).
7 Here Glissant comes back to the longstanding opposition between the notion of 'citizenship' and that of 'barbarism'; other writers have embraced the identity of 'barbarian' as a political and aesthetical framework by overturning its negative connotations in a critical standpoint for decolonial and anti-racist politics. See Yousfi (2022).

References

Abbas, M.A. (1997) *Hong Kong: Culture and the Politics of Disappearance*. Minneapolis: University of Minnesota Press.

Augoyard, J.F. (1979) *Pas à pas. Essai sur le cheminement quotidien en milieu urbain*. Paris: Editions du Seuil.

Azimi, N. (2011) 'Good Intentions'. *Frieze*, 1 March 2011. Available at: https://www.frieze.com/article/good-intentions

Balibar, E. (1994) 'Qu'est-ce qu'une "frontière"?' In *Asile, violence, exclusion en europe. Histoire, analyse, perspective.*, edited by Marie-Claire Caloz-Tschopp, Axel Clevenot, and Maria-Pia Tschopp, 335–343. Genève: Cahiers de la Section des sciences de l'éducation de l'Université de Genève.

Barthes, R. (1985) "Sémiologie et urbanisme." In *L'Aventure sémiologique*. Paris: Seuil.
Benjamin, W. (1982) *Das Passagenwerk*. Frankfurt am Main: Suhrkamp.
Bickford, S. (1996) *The Dissonance of Democracy: Listening, Conflict, and Citizenship*. Ithaca, NY: Cornell University Press.
Biserna, E. (2017) 'SoundBorderscapes. Lending a Critical Ear to the Border'. *AntiAtlas Journal*, no. 2. https://www.antiatlas-journal.net/02-soundborderscapes-lending-a-critical-ear-to-the-border/
Biserna, E. (2021) 'Ambulatory Sound-making. Re-writing, Re-appropriating, "Presencing" Auditory Spaces'. In *Bloomsbury Handbook of Sonic Methodologies*, edited by Michael Bull and Marcel Cobussen, 297–314. New York: Bloomsbury.
Biserna, E. and Raimondo, A. (2022) 'Ascoltare Attraverso. Una conversazione fra Elena Biserna e Anna & Raimondo'. In *L'elettornica è donna. Media, corpi, pratiche transfemministe e queer*, edited by Claudia Attimonelli and Caterina Tomeo, 203–224. Roma: Castelvecchi.
Biserna, E. and Rieussec, C. (2021) Be Water. Soundpocket, Fiona Lee et l'écho des luttes à Hong Kong. *wi watt'heure, Revue & Corrigée* 35. https://www.revue-et-corrigee.net/?v=wwh&a=2021&n=35
Certeau, M. de (1980) *L'invention du quotidien. I Arts de faire*. Paris: Union générale d'éditions.
Corradi Fiumara, G. (1985) *Filosofia dell'ascolto*. Milan: Jaca Book.
Couldry, N. (2010) *Why Voice Matters: Culture and Politics After Neoliberalism*. Los Angeles and London: SAGE.
Debord, G. (1955) 'Introduction à une critique de la géographie urbaine', *Les lèvres nues*, September 1955. Available online: Guy Debord, 'Introduction to a Critique of Urban Geography', accessed 13 September 2021, https://www.cddc.vt.edu/sionline/presitu/geography.html
Drever, J.L. (2020) 'Listening as Methodological Tool: Sounding Soundwalking Methods'. In *The Bloomsbury Handbook of Sonic Methodologies*, edited by Michael Bull and Marcel Cobussen, 599–613. New York City: Bloomsbury Academic.
Glissant, E. (1997) *Poetics of Relation*. Translated by Betsy Wing. Ann Arbor, MI: University of Michigan Press. doi:10.3998/mpub.10257
Hall, S. (1994) 'Cultural Identity and Diaspora'. In *Colonial Discourse and Post-Colonial Theory*, edited byPatrick Williams and Laura Chrisman, 227–237. London: Routledge.
Haraway, D. (1988) 'Situated Knowledges: The Science Question in Feminism and the Privilege of Partial Perspective'. *Feminist Studies* 14(3): 575–599. doi:10.2307/3178066
Ingold, T. (2000) *The Perception of the Environment*. London and New York: Routledge.
IS. (1958) 'Définitions'. *Internationale situationniste*, June 1958, accessed 13 September 2021, https://www.cddc.vt.edu/sionline/presitu/geography.html.http://debordiana.chez.com/francais/is1.htm#definitions
Jacobs, J. (1961) *The Death and Life of Great American Cities*. New York: Random House.
La Cecla, F. (1988) *Perdersi. L'uomo senza ambiente*. Roma: Laterza.
Le Breton, D. (2000) *Eloge de la marche*. Paris: Ed. Métailié.
McCartney, A. (2005) 'Performing Soundwalks for Journées Sonores, Canal de Lachine', in *Performing Nature: Explorations in Ecology and the Arts*, ed. Gabriella Giannachi and Nigel Stewart, Bern: Peter Lang, p. 221.
McCartney, A. (2010) 'Soundwalking and Improvisation'. Improvisation, Community and Social Practice Website. http://www.improvcommunity.ca/
Mezzadra, S. and Neilson, B. (2013) *Border as Method, or, the Multiplication of Labor*. Durham: Duke University Press.
O'Neill, M. and Ismail E. (2019) 'Walking Borders, Risk and Belonging'. *Journal of Public Pedagogies*, no. 4 (November). doi:10.15209/jpp.1173

Paquette, D. and McCartney, A. (2012) 'Soundwalking and the Bodily Exploration of Places'. *Canadian Journal of Communication* 37(1). doi:10.22230/cjc.2012v37n1a2543
Paquot, T. (2008) Dossier "Marcher." *Urbanisme*, 359.
Popescu, G. (2013) 'Re-Visualizing Border Cartographies through Art: From Lines to Flows and Portals'. *AntiAtlas of Borders* (blog), 28 September 2013. https://www.antiatlas.net/re-visualizing-border-cartographies-through-art-from-lines-to-flows-and-portals-gabriel-popescu-en/
Popham, P. (2004) 'The Last Wall Dividing East and West Comes Down'. *The Independent*. 13 February 2004. https://www.independent.co.uk/news/world/europe/the-last-wall-dividing-east-and-west-comes-down-5354855.html
Pötzsch, H. and Brambilla, Ch. (2015) 'Iborder, Borderscapes, Bordering: A Discussion between Holger Pötzsch and Chiara Brambilla'. March 2015. https://www.societyandspace.org/articles/iborder-borderscapes-bordering
Preciado, P.B. (2020) 'Learning from The Virus'. *Artforum*, June 2020. https://www.artforum.com/print/202005/paul-b-preciado-82823
Sadler, S. (1999) *The Situationist City*. Cambridge, MA: The MIT Press.
Sansot, P. (2000) *Chemins aux vents*. Paris: Payot & Rivages.
Smolicki, J. (2021) 'Soundscapes as Archives', *Seismograf*, https://seismograf.org/node/19500
Solnit, R. (2001) *Wanderlust: A History of Walking*. New edition. New York: Penguin Books.
Strüver, A. (2005) *Stories of the 'Boring Border': The Dutch-German Borderscape in People's Minds*. Münster: LIT Verlag.
Thibaud, J.P. (2008) 'Je, Tu, Il. La marche aux trois personnes'. *Urbanisme*, 359: 63–65. Accessed 20 March 2014: hal.archives-ouvertes.fr/docs/00/50/25/89/.../2008_JP_T_ART_Trois.pdf
Thomas, R., ed. (2010) *Marcher en ville. Faire corps, prendre corps, donner corps aux ambiances urbaines*. Paris: Éditions des archives contemporaines.
Truax, B. (1978) *Handbook for Acoustic Ecology*. Vancouver: ARC Publications.
Ultra-red. (2014) *Five Protocols for Organized Listening*. Berlin: Koenig.
Voegelin, S. (2019) *The Political Possibility of Sound: Fragments of Listening*. New York and London: Bloomsbury Academic.
Yousfi, L. (2022) *Rester barbare*. Paris: La Fabrique.

4

THROWING STONES AT NOTHING

Patrick Farmer

Introduction

During the research and writing of this chapter,[1] I fell over numerous times. As a child I fell down the stairs so often that my mother stopped coming to see if I was okay. Most of my childhood memories revolve around falling, an event that has leached into my adult life, the debilitating and mirrored sting I feel whenever I see anyone fall over. It's interesting to think about the ever present threat of falling whilst walking and how over the years it has changed not only what I listen to, but how I listen and what I think I mean by listening.

'Throwing Stones at Nothing' is a piece of fecund speculation, but also of experience, moving away from, as Lynn Margulis and Dorion Sagan put it, a myth of the 'completely self-contained individual' (Margulis and Sagan, 2002, p. 19), in the hope of helping to replace it with a more flexible description. This chapter is implicated in the myriad possible levels of emergence and incorporation in which the composite ego, the self, can come apart, de-personalise, resonate, integrate, drift and reintegrate. In this case, such states are felt and considered through soundwalking, or what I call in relation to my own experience of such a practise, *vertego*. This is a term that relates to ambiguity of space created by the distention of place (a location you can't tell you're 'in' until you perceive it from a distance), one that I understand to merge, block, scatter and correspond in between possibilities of walking and listening; falling and shaking; rushing and splitting.

I was born without the superior layer of the semicircular canal in my right ear, a condition called superior semicircular canal dehiscence. The canal is part of the vestibular capacity of the inner labyrinth. I say this in part because I'm still trying to understand what exactly this means both to, and for, me.[2] I will continue to try to refine the term vertego as and after this chapter unfolds, but

DOI: 10.4324/9781003193135-5

for now is best left as a kinetic happening, a kind of shock wherein ego is ejected from ego (a process I have tried to explain throughout this chapter in a variety of ways), which has led to open and ongoing questions, such as: What returns to ego? What arrives as identity in the motion of movement so often unaware of its own instability? What might some of the differences be between arrival and departure?[3]

According to Margulis and Sagan, 'visible organisms have multiple genomes, if somebody feels like they're falling apart, it's because, on the very level of being, they are falling apart, composed as they are of many different living parts that can be detected and identified' (2002, p. 19). Reflecting on both the abovementioned considerations of the motile ego, and what in this instance I would term the bio-philosophy of Margulis and Sagan, I have attempted to relate and amplify the causes and effects of the vertiginous ejection of ego in my dehiscent right ear to both environment and body–mind.

It follows that the various sections of this chapter can be read in any order. Bar perhaps the Introduction and Conclusion, there is no set route or trajectory. It follows, though not in a straight line, that this chapter is not an attempt to delineate or indeed eradicate boundaries of self; such considerations have arisen in light of the self-reflexive process of soundwalking as vertego as itself pertaining to something other than the self-contained individual. As an aggregate of such bewildered experience, I will consider vertego as reflected in studies of by-products, such things that are 'cast away', in the words of psychologist Andrew Peto,[4] to whom I will, ironically, return.

In sections 6 and 7, this leads to 'Asterion' (the name, among many other things, of the ancient Cretan minotaur), here considered as a complex and drifting mist of symbols based equally on readings of mythological, alchemical and hermetic histories, as much as the inseparable practices of listening and walking, made so by my lived experience of the vestibular and auditory conditions, semicircular canal dehiscence and autophony. By gathering together a mandala of symbols, concepts and metaphors, such as botanical dehiscence, bewilderment, bull, labyrinth, aleph and imagination, this chapter ties and unties soundwalking in a way that, I hope, will present some alternative ways to interact with such a practice as a form of auditory and vestibular cohabitation.

1.

Due to my dehiscence, over the last ten years I've found it increasingly difficult to take part in any group soundwalks. Being in a situation whereupon I stop and listen with others in relation to whatever is happening at that moment (if someone suddenly says, 'look up!', or during a silent group soundwalk, people all look up at a certain spot, or I have to try to keep in step with others as we both walk and listen) has become an increasingly precarious, difficult, and sometimes embarrassing, venture.[5]

For me, soundwalking could be defined as walking and listening in cooperation. If such acts don't cooperate, due, in this instance, to dehiscence, for me it then becomes a matter of finding ways to help them, and by extension myself, to do so, or looking again at what I had thought was competition, which we might here call struggle, to see if it can be otherwise. This in turn has led to a revaluation of such things, not as superficial dichotomisations, but as part of a generatively alchemical coincidence and reciprocity of opposites.

There are myriad ways of holding such multiplicitous processes as competition and cooperation together, holding them to the extent that neither one dominates the other, but simply moves and continues to move, to correspond, creating and suffusing their relations whilst also remaining mindful of what may emerge. The Roman poet Lucretius, and his understanding of what he termed simulacra, seems a good point at which to briefly enter and then leave. For Lucretius (2007), the simulacrum (which I understand to be the energy, the heated breath of the world that rises and falls, departs and returns, prone to all sorts of pareidolia) concealed the shock and motion through which it was produced, lending credence to sensible and perceptual stability, and yet it could also stimulate perceptual, if not sensible, inference, enabling a register in which the world is held together whilst coming apart. Going into this chapter, the presence of Lucretian simulacra will be implicit, joining and separating cooperation and competition, in order to see what else might emerge, but also to pay attention to what I'd missed, and what was already there.

This has led to the development of a vocabulary premised on flux and permeation, on the self-contained and wildly spooling individual, on energy as matter and matter as energy. A vocabulary that walks with language (as a language that walks). It follows that walking is not here premised on a hierarchy of tacit expectations of ability, with regard to gate, posture, tone … but to assemblage and slow movement; to such things as are thought to reside in (perhaps I should also say return to) the vestibular labyrinth of the inner ear, a place that certain psychologists believe coevolved with both the emergence, development and drift of the superego in infants (Peto, 1970, p. 401), to which, once again, I will return.

Pertaining to the self-contained nature of an individual, Margulis and Sagan ask, 'when the committee [here they are referring to the notion that organisms are walking assemblages] gets sick, is simply a single animal getting sick, or is "illness" more a rearrangement of the members?' (2002, p. 19). For me, this question is one way of trying to understand the unshakeable feelings and images I receive during periods of vertego. The only place into which I can withdraw during such periods of vertiginous disturbance is a movement in place (a walk) in which I have to try to not try to listen. This is often reflected in the fact that, when I'm overly conscious of the rhythm of my walking, an imbalance emerges through the very effort of awareness. Does such a disturbance shake up the whole of me, or part of me? If it is the former, or can be, what then happens to my ability to listen? Let alone to walk. And indeed, if it is the whole of me that is shaken up, that comes apart, does it follow that listening and walking are also rearranged?

I am aware of (perhaps even impishly enjoying) both playing with and leading on the irony that in order to write about my selves falling apart and casting away, I also have to write about myself as if I were not, as if I were indeed, a self-contained self. This is a still-shaking of self that can indeed be very painful, but it can also present other ways to think about processes of individuation, about relations among living parts, ratios of change, the curious eventuality of being able to listen with an other degree of clarity (far from solely aural), as I am being shaken up. The cultivation of the notion and experience of vertego lends itself to an ontology akin to an alchemical chord wherein such terms as destabilise and stabilise, in reference to the locations of the ever-shifting boundaries of self, also seem to become tired in their use.

2.

In Charles Darwin's *The Power of Movement in Plants*, published in 1880, at least a third of the pages are taken up with drawings denoting the circumnutation of seedlings. These drawings, which I believe were made by Darwin, remind me very much of spikes of vertego, my quavering ability to walk when listening (this refers to my entire body) anywhere but front-centre (otherwise known as the Sagittal plane) (Figure 4.1).

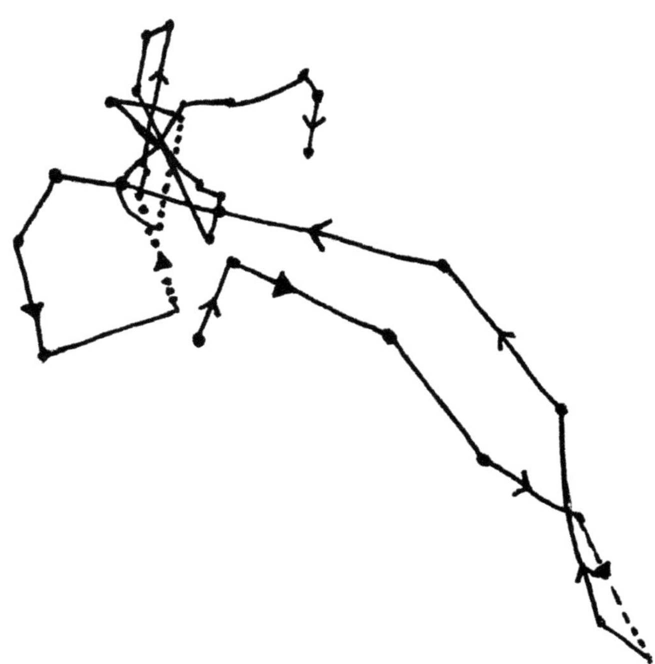

FIGURE 4.1 The movement of plants. Illustration by Patrick Farmer, based on Charles Darwin's notes on plant movement

Circumnutation is a heliacal organ movement widespread among plants and is influenced by light, touch, temperature, sound, chemicals and of course, gravity. Curiously, sessility, in relation to motility, refers to an organism's means of self-locomotion, however, in a botanical sense, sessility means 'sitting', referring to plant parts that have no stalk. Botanically, dehiscence is known as the splitting along a so-called 'line of weakness' in a plant; this is in order to release, sometimes quite violently, its contents. Certain plants can fling their seeds over 100 m, creating quite remarkable cracking sounds in the process. Dehiscence is often thought of as a process of abscission, which is to say a shedding of body parts, with the ejected structures said to be to caducous, which means 'falling'. Anatomically speaking, in any creature with semicircular canals, dehiscence is most often used to refer to a 'surgical complication', where the edges of a wound no longer meet. Superior semicircular canal dehiscence, first reported by Lloyd B. Minor in the 1990s (Minor et al., 1998, p. 249) refers to a thinning, or a complete absence, of the temporal bone that overlies that part of the superior semicircular canal, and is often congenital.

The human inner ear is 'usually' a closed capsule; there is only one way 'in', and one way 'out'. Vibration is transmuted through the stapes bone into the inner ear, which then distils (as pressure) through the round window. This whole process causes the oval window to bulge in, as the round window bulbs out. Between the balance and hearing organs there is a kind of berm, a cartilaginous embankment. Semicircular canal dehiscence can result in what we might call a third window (a hole into and out of the inner ear) in addition to the round and oval windows. This allows motion of the endolymph, a conductive fluid, to be induced by both sound and balance stimuli. It essentially enables vibration and electricity to flow over and through the boundary.

The hole between the inner ear and the brain cavity can be as small as the tip of a ballpoint pen; though it's important to note that there are multiple bores along the plane of the bone, much like the exit holes made by beetles through the trunk of a tree. Instead of spiralling throughout the cochlea, some of the vibration that passes into the inner ear is ejected out of the third hole of the canal. This means that when subsequent fluid moves, 'unnaturally', let's cautiously say, in the balance organs, the brain receives afferent signals, as if the head were moving, though for all intents and purposes, it is not.

But sound is part of bone, a body is in so many ways the sound of its own integration. Pulse, heart, feet; the hiss and static of muscles, billions upon billions of biochemical responses … These energies can penetrate the balance organ of the inner ear through the third hole, and cause havoc; projecting self into a space of impossibility, one where it feels as if there are as many definitions and experiences of sound as there are sounds in the universe.

At the life of this hole, sound is in essence made and remade, in much the same way as the brain can pull apart and remake a hydrogen atom. If we think of the human voice, for example, as it travels out of the mouth and around to the ears, it also travels through the jawbone and passes into the hole, disturbing the balance organs, aggravating the voice as if at least two people were speaking, creating dendritic time and sensation. During such moments, the sounds I experience coalesce into a spaceless echo of my bodies, as I don't think they pass through the spiral of the cochlea, but rather reflect and ricochet around bone and the brain cavity. I think that what I feel during such moments is the sound of vibration itself.

Dehiscence turns my legs to jelly, a feeling comparable to sudden and unexplainable spikes of anxiety. Autophony, a side effect of dehiscence which lifts my footsteps into my skull and back, unavoidably turns and folds, listening up against itself. This is the emotion of motion in which my eyes, tongue, jaw, heart and bones absorb any residue of electromagnetism, accentuating the texture of the vibrating surface, a distension that treats the body as if it were a bellows.

3.

Vertego doesn't answer questions so much as lengthen the resonance of questions unheard. It is to spin whilst passing oneself on the way round. Such a bewildered self is more akin to feedback than hierarchy, which is to say, each description of an I is implied by another and in turn implies, or uncovers, another. To bewilder in this sense is to lose sense of where one is, where one hears, parts folding in the midst of unfolding, a nesting of larger in smaller holes. For the poet Fanny Howe, bewilderment offers a walk into a wild place, on 'the threshold of love's sanctuary which lies above reason' (Howe, 2003, p. 16). Another side of this experience is one that ejects preconception and normativity in favour of openness and experimentation.

A spiral can be an experience of perplexity and loss of bearing. When walking is full of falling, stumbling and pain. Trying to get in deeper and to get safe at the same time. Progressing at one level and becoming more and more lost at another. In this sense, to balance is also to listen, a subtly conjoined attempt to stabilise flux, or the opposite, to wrest one thing from another and watch them turn. For such a spiral walker there is no plain path, no up and down, no inside or outside. It is to walk like sound walks, the asymmetry of empty space in which there are strange returns and recognitions, but rarely a conclusion. What goes in, at that time, goes out. The feeling that I'm full, metabolically and energetically speaking, leads to the question, how can I empty myself out? And the only way I've ever known how to do that, is walking. Each movement forwards catches and diffracts what is coming toward, which of course is also away. Intention reverses along attention. Spinning is central to centreless bewilderment. An approach, dizzying and repetitive, dissolving particularities into a solid hollow of sound, matching the outwardly observed with the internally heard.

These subtle contingencies of obliteration are the worlds that encircle space, measurements beyond correlation. Howe developed her appreciation of bewilderment by assiduously living in the world, ejecting attention out and into her poems, rushing back and forth within an irreconcilable set of imperatives. Walking to empty leads through such points of repose which I equate to unlistening (by which I mean, not locating, walking in order to try not to try to listen, listening in order to try not to try focusing on walking), wherein listening-walking-selves have split apart, recombined, ejected.

4.

The human vestibular apparatus is not only an organ for perception (of movements and gravity), but gives rise to reflexes, to turning and progressive movement, in turn influencing muscle tone, gate and posture. Sensations from the vestibular nerve do not necessarily form a part of our conscious knowledge of the world. Such a sensory organ, full of preconscious impressions, leads to labyrinthine motility of an instinctive life, highly sensitive to emotions and the development of both imagination and psyche, where changes in either can succinctly express themselves in vestibular sensations, and in consciousness.

It may seem like a prematurely concluded individual who can stand apart from such experiences as I have been describing. In truth, the inclination, as aside from the ability, which quavers in intent and capability, is one that wishes to communicate, to transubstantiate the experience of what feels like an ever-widening gulf, in which the pain and turmoil that accompany the loss (a plurality of states all drawn toward that roving centre) integrate through their opposites in order to be heard and felt.

In the 1930s, psychologist Paul Ferdinand Schilder devised a list rooted in such affective observation and experience, a list that focused on the influence of gravity on emotions developed in accord with his understanding of the vestibular labyrinth of the human inner ear (Schilder, 1970, p. 94; 1933). Schilder would write as he simultaneously rode an elevator, assiduously monitoring his reactions with an intense ardour.

Schilder claimed that everyone has a potentially infinite number of body-images, and that the human psyche is a physical phenomenon, suspended in such space that we are affectively moving through as we are being moved, a kind of gravitationally concentric wax. 'When the elevator is stopped,' Schilder states, 'the legs become heavier, but the rest of the body continues to go down, so that below the feet there are two lighter phantom feet' (1970, p. 95). He felt that vestibular irritation must change the perception of the gravity of our own bodies, so he studied the roles of gravity as a linear accelerative force, one which constantly influences vestibular function in connection with human development and behaviour.

In the 1980s, psychiatrist David Hubbard astutely pointed out that Schilder's observations of the ego and its perceptions prevented him from riding in elevators in an absent-minded way as so many of us do (Hubbard and Wright, 1985, p. 162). A notational sense of being shifted within and without his body as the elevator upon which he rode accelerated in relation to gravity. Clearly, as that which is inherently vestibular is irritated (due to excitation in the sphere of the otoliths and the planes of the semicircular canals), parts of the substance of the body–mind wander, creating what we might call simulacra of indistinct and inaudible in-outlines.

'The emanation of the substance of the head out of its frame is of special importance,' writes Schilder in a beautifully Neoplatonic fashion (1970, p. 96). He goes on to state that this emanating substance is the carrier of the localisation of the ego, nominally located in the height of the frontal bone between the eyes. And yet, when there is vestibular after-sensation, that which trembles becomes the carrier of the ego, a kind of lengthening and shortening of the body, a temporal dissociation between 'vestibular' head and 'real' legs.

Considering that the ego could be located between the eyes cannot but lead me to think of D.W. Winnicott's theory of 'ego orgasm', so beautifully espoused by Catherine Clement, in which the ego cannot be localised, as 'it invades the body, the skin, the consciousness' (Clement, 1994, p. 217). Such a point of view cultivates an image of deformation, of body–mind–environment always at a remove, each ejecting from, and subsequently merging with, another location.

The movement between this section and the next is characterised by the proposition that gravity itself holds and manifests emotion in a manner familiar to the multiplicitous individual, permeating the boundaries of identity and ensouling the world. As the philosopher Martin Buber writes, 'everything broken points to the unbroken' (Howe, 2003, p. 13), to which Fanny Howe responds, 'the awareness of both continuum and rupture occurring together may form the very rhythm of consciousness' (2003, p. 13).

5.

If to be bewildered is to lose sight of where we are, to forget where we've been, to be cast away, what if we didn't know where we were to begin with? Bewilderment can follow a collapse of reference and reconcilability, cracking open dialectics to witness the motion of myriad selves.

Writing in the 1970s, psychologist Andrew Peto claimed that the vestibular system acts as a precursor of those mental functions which are grouped together as superego: 'This vestibular forerunner includes imagery and fantasies which develop in the course of those traumatic experiences of the child that are part of normal maturational and developmental conflicts' (1970, p. 401). Peto writes through Schilder and the stresses he laid on disturbances of vestibular life in relation to depersonalisation, or the equilibrium of disturbance and interaction between ego and environment, claiming, in a manner that echoes Margulis and Sagan, that whenever a person is giddy they cannot maintain the unity of their body (1970, p. 408). Peto observed that such situations recreate childhood feelings of being abandoned and losing one's balance, wherein vestibular interplay between the superego and the ego can be perceived as physical disorientation and giddiness.

'A situation of being cast away comes about and giddiness develops, symbolising in a somatic regressive form the moral conflict between ego and superego' (Peto, 1970, p. 401). This feeling may be experienced in a very concrete way, manifesting in confusion, despair, bewilderment, 'one part of the self may look at the other part or may feel in some eery way that it is separated from the other part' (Peto, 1970, p. 407).

Building on the research of Peto and Schilder, David Hubbard and Charles Wright studied the potential influence of gravity in relation to human personality development, and further speculated that, alongside the more traditional stages of psychosexual development, those of oral, anal and genital, alongside the phallic and latent, there existed the gravitational (operative over a lengthy period of personality development), realising that it was in dire need of deeper consideration and attention (Hubbard and Wright, 1985, p. 162). Hubbard and Wright stipulated that such a gravitational stage could be more influential than many of the aforementioned Freudian stages of development, particularly in relation to personality function and neurological reflex. Subsequently they returned to the image of the infant's first years of life, learning to move, to orient their physical body against and with such an ever-present force as gravity. Not to mention the subsequent maturational life of motor skill development and coordination, in spite of such a pervasive energy, wherein the fear of gravitational pull may well serve as the paradigm of all subsequent fears.

Evidently, the attempt to overcome gravity, to become with gravity, constitutes a principle struggle throughout the early years of existence. The child, if possible, manages to control their position within the crib, to sit, crawl, walk, run, jump, fall, even to be still. Gravity, in this sense, is felt to be the cause of the first ego strain, leading Freud to remark, 'Where id was, ego shall be' (Freud, 1933, p. 112), to which Hubbard replied, 'and its perpetual antagonist is gravity, which speaks through its own organ' (Hubbard, 1971, p. 72).

Conscious sensation and affect might arise not only from motion, but from loss or gain of otoconial mass. The one is the other, in a way ... The acuity of the hearing ear can be accurate in terms of one-millionth of a second in the localisation of a sound source, and it is conceivable, according to Hubbard, 'that the phylogenetically older vestibular system could have equal or even greater sensitivity in the perception of stimulus variations' (Hubbard and Wright, 1985, p. 181).

In his paper on the 'auditory sphere', Otto Isakower (1938, p. 340) begins by commenting on Josef Breuer's suggestion (in the late 1800s) that the otolith organs do not play a role in the function of hearing, rather in the perception of the movement and position of the body, relative to environment. Isakower, a physician, relays an intriguing observation. He tells us that certain crustaceans lose their otoliths, together with their shell, at every moult, subsequently developing new ones, often by searching the aquarium floor, in order that they might fill their otocysts[6] with grains of sand, literally coming apart, casting away. Ingeniously, Isakower compares this process of integration to the development of the ego, part of which he calls the human auditory sphere (1939, pp. 341–342). The superego is held then, as a mental derivative of a separate force, that of gravity, of environment, functioning, as Peto states, like a 'psychical organ of equilibrium' (1970, p. 402).

6.

In *Software for People*, Pauline Oliveros writes that she would like to amplify the sound of a bull dozing (1984, p. 18). In my ear, I hear a bull rushing. Bulrushes disperse their seeds and spores via a type of dehiscence, producing millions of dust-like grains that travel on the wind, the beaks of birds, human feet, and much more besides, moving in a manner similar to how I imagine vestibular irritations, particularly when I bring to mind Peto's claim that 'whenever a person is giddy, they cannot maintain the unity of their body' (1970, p. 408).

The following sections are an attempt to mix together being and doing with images of vertego (the 'individual' ever coming apart as together), corresponding to a creature both 'imaginal' and 'real'. The wild and symbolic nature of the bull goes hand in hand with the idea of a labyrinth, an overriding purpose and idea of being lost. Asterion is the name of the Cretan minotaur of the labyrinth, previously a dance floor, built by the inventor Daedalus. In human anatomy, the Asterion is a visible point on the exposed skull, just behind the ear. The very presence of Asterion symbolises the enticement of a hole, a pulse where three cranial bones meet. Many neurosurgeons use the Asterion to orientate themselves in order to plot safe entry into the skull. Coins minted in Knossos from the fifth century show the bull, kneeling, adorned with cereals and grains; the underside shows a star at the centre of a series of meander patterns. The star is itself, Asterion, part of the constellation, Canes Venatici. Bull emerges from itself, and thus is more than it appears.

Many representations of the bull push it into the ring or the farm, a soulless machine,[7] but the bull that is Asterion still wanders the labyrinth. Reading 'The House of Asterion', a short story by Jorge Luis Borges, told from the point of view of the bull (Borges, 2000a, pp. 170–173), helped me to realise that the dehiscent ejection of ego is at the same time akin to receiving what returns, that the outside always leaks in, creating an indefinite and perpetual vacillation of egos.

Brownian motion, otherwise known as pedesis, or 'leaping', describes the random movements of particles, detailing how larger particles can be moved by the impact of many smaller masses.[8] It is a visible means by which to detect invisible effects, a discontinuous structure of matter that appears to be 'a one-dimensional random walk' (Knight, 1962, p. 218) that can sometimes look remarkably like the diagrams in Darwin's book concerning plant motion. It can even help one imagine just how Asterion runs and walks in the Borgesian labyrinth, the lines of which, as Borges himself states, trace the lineaments of his own face (Borges, 2000b, p. 183) (Figure 4.2).

FIGURE 4.2 Taking an atom for a walk. Illustration by Patrick Farmer, based on Jean Perrin's notes on atomic motion

Regarding my predilection for picturing bull in terms of Brownian motion, I don't so much concentrate on 'seeing' bull the way in which I picture them, but in the light by which I see them. In this way, bull is not a colour, not a form, but the phenomenal means by which I am able to see in light, hear in sound, etc. Bull is a labyrinth roaming a labyrinth, ruminating, a spinning structure of convulsive and ejecting meaning, a living chronology of fractal shifts and unpredictable causality. Tiny rootlets of chiral vibration fill such inner wind like a meshwork of flashing interstices, dendritic roots detach themselves with vertiginous speed from the wind-contracted mass that swallows them.

Bull could then be another term for an ego that is in the self-same process of returning and ejecting, a monstrous multiplicity that distends, parts and reunites with its roaming indefinite individuality. In 1883 the mathematician Georg Cantor created what he called a 'monster', taking a straight line and breaking it into thirds, erasing the middle third, taking those two lines, taking out the middle thirds, and doing it again and again and again, spiraling through infinite points where, under scrutiny, the Cantor set (otherwise known as fractal geometry), appeared to stay the same (see Jersey and Schwarz, 2008). In a manner familiar to this infinite identity of the splitting stillness of appearance, bull rushes in an ancient labyrinth that is ever without emptiness, regardless of how it may appear. Within the ear we might find, on the level of its energy, a glimmer that is the reflection of the shedding of magnitude, a camera silenta[9] in which bodies within bodies transform as ruminants of each other's edges.

Ego without ego, passing through and returning in maturational stages, never seems to reach fructification. Sound shakes into patterns and bull becomes a star, a harp of periodic harmony and dissonance at the centre of a centreless labyrinth. As subjectivity dissolves more is revealed, akin to the monads of Leibniz, where each portion of matter may be conceived as 'a garden full of plants, and as a pond full of fish' (Leibniz, 1960, p. 466). Each branch of every plant, each member of every animal, each drop of their liquid parts, is in itself a self-similar garden or pond in a philosophy of organism, wholes that are made up of parts that are themselves wholes on another level. This is pollen moving on the surface of a lake that is itself a layer in an organism of revolving life, an indefinite variation of forms, of patterns, of insights, the minotaur as minted coin, the energy drawn from constellations.

Everywhere around the ear such energy swirls in the midst of its own transduction, bending motions in spiral volumes, producing affective movement in and of its extension. Like Isakower's crustaceans and the goal of Margulis and Sagan to shake off competition and cooperation, parts of self and environment are constantly recombined, enfolding in development rather than pitting one thing against another, introjected into integral parts of orienting elements and bodies, or what Isakower calls an 'auditory sphere'.[10]

7.

As depth psychologist James Hillman reminds us, the term 'animal spirits' was once used in psychological explanations used throughout Western history. It referred to 'what goes on inside' (Hillman, 2008, p. 61). The first letter of the Hebrew alphabet, aleph, derives from a glyph that represents the face of an ox, or bull. A is not then for apple and the fall, A is for aleph and bull.

The aleph is a fourth dimension, one that captivated Borges as a child, playing with a collection of four-dimensional cubes, composing a bright tesseract, an infinite combination of the planes that nevertheless cannot be reduced to any sense of their supposed totality (Couliano, 1991, p. 27). Just as Schilder felt psyche to be a physical substance, the same could be said of the aleph, an analogue of a fourth dimension which would require, in this context, a fourth canal in the ear, a hypercanal, vibrating to such a fine degree that it is pure energy, the electricity that drives gravity.

Hillman writes that, 'when confronted with a great animal power, we do not sing to it, bow down, dance in its honour, rather, we interpret it. We wave the hermeneutic wand to make the force go away by transforming it into a meaning' (2008, p. 62). In other words, we so often meet bull with our own bull. The power of the image of bull, says Hillman, 'is like the power of the imagination itself' (2008, p. 63), Martial and Venusian vibration, earthbound and cosmogonic. Bull is another way of channelling speculative fecundity, the ways in which the ecstatic moves out and the enstatic moves in. Borges' Asterion longs for ears that hear all the sounds of the world (Borges, 2000a, p. 172). So what, I wonder, might such appendages look like?

Just As Borges' Asterion seeks to hear all sound, the aleph is said to be the source of all articulate sound, a perfectly spherical zero, akin to disintegration of the ego, altering balance and orientation. 'To hear the aleph,' says Gershom Scholem, 'is to hear next to nothing' (1965, p. 30). It is the preparation for all audible language, but in itself conveys no determinate, specific meaning within an infinite system of meaning devoid of specific meaning, always on the verge of translation, of doubling, of splitting.

A spectre haunts the 'Aleph', another short story by Borges that draws from infinity the possibilities of cacophony and chaos. Its protagonist, one of Borges' innumerable alter egos (in which those egos too possess an alter ego, and so on)[11] mourns the death of a lost love, ruminates on the impossibility of stone in which 'the universe shall change, but I shall not' (Borges, 2000b, p. 118). Reading and spending time with Borges' 'Aleph', or the immanent situations of the Leibnizian Monad, I realise again and again that I don't know how I listen, but I do know how I imagine listening, particularly as I walk. Vertego is a means of participating in such dissipation, where coming to know a world is indelibly linked to changing it and being changed by it. Dehiscence leads me, as gravity leads it, along undulating and spherical degrees of the vertiginous and auditory imaginal, never more active than when I am walking, when I am, with every footstep, ejecting and returning among a vibrating field of relations.

Conclusion

If we think back to the hole in the canal of the ear's labyrinth, we imagine bull, Asterion, as a vibration within a vibration, and so on. Inevitable step by obstructive step, infinitesimal vortices in a maelstrom of sinusoidal tones. I imagine bull as the constant growth of bodies dispersed in infinite divisions of matter and energy, sucked in and out of the labyrinth via the recapitulant third window, the hole that is also a fold.

A reaction need not always come directly from a subsequent action, and so the aleph, a simultaneous occupation and evacuation of all things, presents a shift in the micromolecular structure of the semicircular canal, which in fact, may not register in the eye, as it has been shown to do (Branona and Straka, 2018, p. 1) but in an entangled wave passing through the feathers of a migratory bird 20 years in the future. This is in part how I have come to consider listening as by-product, the world drawn into the body that is in the world, an undulating flow of temporal dissipation wherein listening is both expulsion of energy and return of matter.

Borges describes how an aleph may exist in the heart of one of the stone columns surrounding the central courtyard of the Amr Mosque in Cairo, inside of which, he says, lies the universe (Borges, 2000, p. 133). This is the vestige of the primal sea that remains in chemical reactions linking the mitochondria in our cells to celestial cytoplasm, the illusion of irreversibility intimately affected by us, and yet it is one our eyes cannot see. The stone avails itself in bustling rumour to the ear that is willing to return to the ground, to both shed and retain its bipedal imagination, like the ipecac ego, both milking and growing its auditory lobe.[12]

The Borgesian aleph (which we could almost think of as an archetypal dissipative structure, the memories of atoms held apart by gravity, the frequencies buried in those in between spaces) maintains its identity only by remaining open to the flux of environment, by equating timelessness to time. I have tried in a variety of ways to picture, imagine, fabulate and communicate vertegonous self, a stratified and emergent gravitational vacillation, one that retains its structure because it is at the same time falling apart, a relational process of patterning in which to listen is to be listened to, to walk is to be walked (Figure 4.3). Walking with the influence of vertego, as if it were a constellation, these differences can also be felt in reverse. Gravity breaks the cast of the self that is beyond the control of the ego and the anatomical planes, and so dehiscence does not survive the difference of inside and outside, of listening and walking, or is it the other way around?

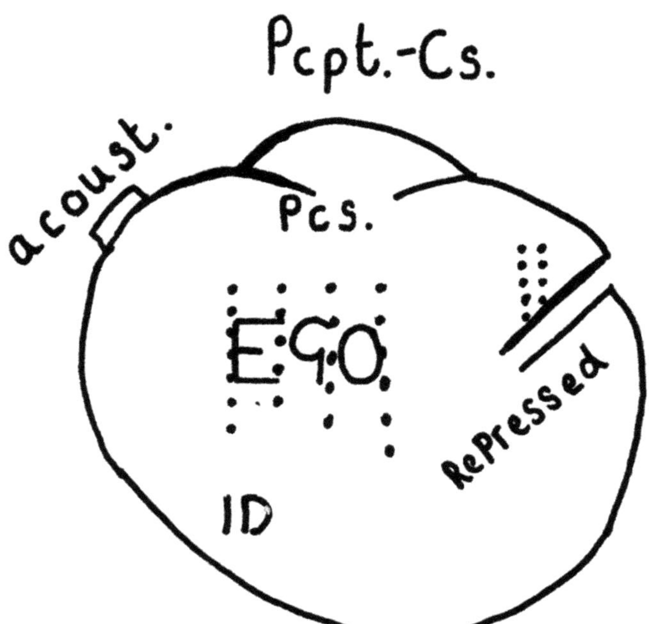

FIGURE 4.3 An auditory lobe. Illustration by Patrick Farmer, based on Sigmund Freud's structural model of the mind

The vacillating and fuzzy vantage point that is semicircular canal dehiscence is for me a threshold of unlistening. Just as sound walks, vertego draws a line only to try and cross it as I try not to try. It is an ejection that is somehow also a return, a multiplicity of auditory relationality, balance and body–mind–environment, that, to paraphrase Margulis one last time, neither competes nor cooperates, but cohabits and straddles.

Notes

1 By necessity I have engaged with certain anatomical details, and in mind of such I have written a brief introduction to the vestibular labyrinth of the human inner ear, which can be found at: https://patrickfarmer.org/notes-on-the-ear/. I have also included a short glossary of terms. Anyone who is interested can find a number of notes, including those on the auditory labyrinth, that have quietly fed into the metapattern formed around the conjunction of a series of essays, of which 'Throwing Stones at Nothing' is a part.

2 Thinking about how anything hears, or indeed does not, is for me a moment-by-moment prediction and interrelationship that will by its very nature alter how we hear and equally what we're hearing, and not. Dehiscence strains and elaborates upon my abilities to walk and to listen, to be with, as, and among the world.

3 Following the psychoanalyst Melanie Klein, in this chapter the terms 'ego' and 'self' are interchangeable. I see them as shifting volutes of positionality in which one serves to differentiate 'me' from 'not-me'. See Hinshelwood (1991).

4 This chapter looks at Peto's theory of a vestibular superego in a variety of mental and physical phenomena (Peto, 1970, pp. 401–416).

5 For more information about the events that led to the exacerbation of these conditions see Farmer (2019).

6 Generally now known as a statocyst, an organ of balance and orientation in many aquatic invertebrates.

7 According to David C. Smith, cows are 40-gallon fermentation tanks of four legs. In contrast to this, Margulis writes that they, cows, 'ingest grass but they never digest it because they are incapable of cellulose breakdown. Digestion in cows is by microbial symbionts in the rumen. The rumen is a special stomach, really an overgrown oesophagus, that has changed over evolutionary time. Cows that lack rumens don't exist; cows (and bulls) deprived of their microbial symbionts are dead.' (Margulis and Sagan, 2002, pp. 14–15).

8 For an overview of this phenomenon, see (Perrin, 1916, pp. 109–133).

9 For those interested in learning more about the development of the silent room, camera silenta and communication technologies, see Schmidgen (2013).

10 For Isakower, the term 'sphere' includes 'not only the whole apparatus in all its portions from the periphery to the centre, but also the part of the psychical structure to which the actual sensory apparatus is subordinated and which provides for the further elaboration of incoming perceptions' (Isakower, 1939, p. 342). I can't help but append to this eerily mechanistic (yet addictively curious) premise, an observation made by Arthur Koestler in his book *The Sleepwalkers*. In a section titled 'The Arguments for the Earth's Motion', Koestler writes on the Copernican orthodoxy regarding circles and spheres, wherein gravity is a 'natural inclination' combining parts into the form of a sphere, 'contributing to their unity and wholeness', a property Copernicus claimed to be present even in the sun. 'Thus the properties of a whole stick together because of their desire to make a perfect shape,' responds Koestler, and thus gravity, to Copernicus, 'is the nostalgia of things to become spheres' (Koestler, 1989, p. 199)

11 A manner of splitting or moulting that bears similarity to the notion that the amplification of bifurcation can lead to order or to chaos, the proverbial garden of forking paths. See (Borges, 2000a, p. 282).

12 In *The Ego and the Id*, published in 1923, Freud perceives the ego has on one side an auditory lobe, quite literally 'a cap of hearing that is worn crooked' (1933, p. 29). In his revision of 1933, the auditory lobe has been replaced by the super ego (Freud, 1933, p. 111). To me this suggests that the superego is, in part, an appendage (in direct proximity to its vestibular counterpart) through which the mutable ego listens to itself whilst attending with the equally mutable and intertwined body as it moves and is moved through the world. This is reflected in Peto's essay where, if the vestibular superego abandons (casts away) the ego, the subsequent loss of balance can cause 'a general transient disorientation of the ego, where it interprets such giddiness as a sign of disapproval by the superego' (Peto, 1970, pp. 406–407).

References

Borges, J.L. (2000a) *Labyrinths*. London: Penguin Classics.

Borges, J.L. (2000b) *The Aleph*. London: Penguin Books.

Branona, F. and Straka, H. (2018) Semicircular Canal Influences on the Developmental Tuning of the Translational Vestibulo-Ocular Reflex. *Frontiers in Neurology*, 9, 1–13.

Clement, C. (1994) *Syncope, the Philosophy of Rapture*. Minneapolis, MN: University of Minnesota Press.

Couliano, I. (1991) *Out of This World*. Boston, MA: Shambala.

Darwin, C. (1880) *The Power of Movement in Plants*. London: John Murray.

Farmer, P. (2019) *Azimuth, the Ecology of an Ear*. Oxford: SARU.

Freud, S. (1933) *New Introductory Lectures on Psycho-Analysis*. New York: Carlton House.

Hillman, J. (2008) *Animal Presences*. Washington, DC: Spring Publications.

Hinshelwood, R.D. (1991) *A Dictionary of Kleinian Thought*. London: Free Association Books.

Howe, F. (2003) *The Wedding Dress*. London: University of California Press.

Hubbard, D.G. (1971) *The Skyjacker*. New York: Macmillan.

Hubbard, D.G., and Wright, C.G. (1985) The Emotion of Motion: Functions of the Vestibular Apparatus. In: D.A. Shaskan and W.L. Roller, eds. *Paul Schilder, Mind Explorer*. New York: Human Sciences Press.

Isakower, O. (1939) On the Exceptional Position of the Auditory Sphere. *Psychoanalytic Quarterly*, 11, 340–348.

Jersey, B., and Schwarz, M. (2008) *Fractals, Hunting the Hidden Dimension* [film]. Virginia: NOVA.

Knight, F.B. (1962) On the Random Walk and Brownian Motion. *Transactions of the American Mathematical Society*, 103(2), 218–228.

Koestler, A. (1989) *The Sleepwalkers: A History of Man's Changing Vision of the Universe*. London: Penguin Books.

Leibniz, G. (1960) *The Rationalists*. New York: Dolphin Books.

Lucretius. (2007) *The Nature of Things*. London: Penguin Classics.

Margulis, L., and Sagan, D. (2002) *Acquiring Genomes: A Theory of the Origins of Species*. New York: Basic Books.

Minor, L.B. Solomon, D., and Zinreich, M.D. (1998) Sound- and/or Pressure-Induced Vertigo Due to Bone Dehiscence of the Superior Semicircular Canal. *Archives of Otolaryngology –Head & Neck Surgery*, 124, 249–258.

Oliveros, P. (1984) *Software for People*. Baltimore, MD: Smith Publications.

Perrin, J. (1916) *Atoms*. London: Constable & Company.
Peto, A. (1970) To Cast Away. A Vestibular Forerunner of the Superego. *The Psychoanalytic Study of the Child*, 25, 401–416.
Schilder, P. (1933) The Vestibular Apparatus in Neurosis and Psychosis. *The Journal of Nervous and Mental Disease*, 78(1/23), 137–164.
Schilder, P. (1970) *The Image and Appearance of the Human Body*. New York: International Universities Press.
Schmidgen, H. (2013) Camera Silenta: Time Experiments, Media Networks, and the Experience of Organlessness. *Osiris*, 28(1).
Scholem, G. (1965) *On the Kabbalah and its Symbolism*. New York: Schocken Books.

Glossary

Otoconial mass The totality of minute crystals, made of calcite, covering the otolithic membrane.

Otolith organs Gravity receptors, the utriculus and sacculus, located inside the vestibular labyrinth.

Semicircular canals Three half-circular interconnected tubes, located inside the vestibular labyrinth.

5

SOUNDWALKING EXTINCTION: LISTENING ON BORROWED TIME

Milena Droumeva

Introduction

In thinking about soundwalking as a "theory" and a practice across academia, art, and activism we must inevitably talk about the ecological crisis. Not only is soundwalking a method directly emergent from the ecological thinking of late-1960s North America, but its very actualization depends upon the existence of an "ecological" environment to listen to: after all, the acoustic ecology movement began with an appreciation for the balanced complexity of natural environments and is predicated on their reproduction within healthy ecologies. Of course, subsequent conceptualizations of soundwalking take on urban complexity as well. However, located in these ethical and political questions is also the issue of listening itself: what does it mean to listen? More specifically, what is the aim, outcome, or "product" of soundwalking? I want to acknowledge at the onset that there are many different methods and personal approaches to soundwalking, and that soundwalking is used in vastly different contexts: as meditation; as an art practice; as performance; as a technique; and as a research method (Gutierrez, 2019; Taylor and Fernström, 2017; Ceraso, 2014). Depending on the context it can also be less or more common for soundwalking practitioners to use technology, including hidden or embedded audio players, tactile mics and sensors, field recorders, mobile devices, and augmented reality.[1] This technological mediation brings with it its own problematics and discourses of authenticity and added value, particularly in relation to environmentally precarious locations (Biserna, 2015; O Keeffe, 2015a; Droumeva, 2017). My own positionality locates me in an academic context, where soundwalking is a method for ear training, placemaking, exploring urban issues, mapping, and generally an approach to evaluating soundscapes. I use soundwalking as a way of putting participants into a state of attentiveness to sound: a starting point towards further discussion about urban livability and subjective experience. Soundwalking is also my

DOI: 10.4324/9781003193135-6

main personal methodology as a scholar and researcher for keeping myself immersed in sonic inquiry (Droumeva and Jordan, 2019). Performing soundwalks as part of sound ethnography work often leads me to use field recording as a complementary "data gathering" activity. My practice over the past decade has me increasingly reflecting on the ethics of soundwalking and field recording both in urban and in natural settings.

Some of the most notable acoustic ecologists are life-long recordists of rapidly changing (previously rich) natural habitats: from the South American, to the Australian rainforests, the Arctic, and California's redwoods forests, and the African savannah (Krause, 2012; Monacchi, 2014). While these recordings are often touted as artefacts for raising awareness, it is now time to ask, what next? People have been aware of climate change and species extinction for a long time. As sound artists, as field recordists, as acoustic ecologists, we cannot simply continue to memorialize a pristine nature that is disappearing before our ears. Nowhere is that urgency more sharply felt, I think, than in the practice of soundwalking. The combination of it being a personal, reflexive practice, but also a conspicuous and voyeuristic activity (especially when done in groups), makes one painfully aware of soundwalking through an anthropocenic world: a world unconcerned with the extinction of the very habitats we arrive to listen to. What methods are appropriate then—how do we redefine the practice of soundwalking—with a notion of beyond-human ethic, and mobilities justice (Scott, 2020)? If we can't soundwalk to combat climate change, how do we soundwalk *in* climate change? Eco-feminism already sets out some ethical parameters for realistic forms of engaging with our troubled world, through forms of kinship and care (Haraway, 2016). Along with a brief history of soundwalking as research method, I offer in this chapter three auditory scenes of soundwalking extinction: scenes that are indicative, I suggest, of the types of existential and ethical questions that the acoustic community is wrestling with at this moment in time.

Some history of soundwalking in acoustic ecology

In the decades since the emergence of acoustic ecology as a movement and a field, soundwalking has come to possess almost mystical qualities. Listening is often promised to bring us into a special relationship with the world, to reveal hidden qualities, truths, and aspects of reality (Voegelin, 2014). In Hildegard Westerkamp's (2001/1974) original definition, soundwalking is an excursion for the ears and "its focus is to rediscover and reactivate our sense of hearing." Westerkamp suggests that by consciously paying attention to our soundscape we can begin to observe shifts in our relationship to our sonic environment. Similarly, in some of her early work with soundwalking, Andra McCartney[2] notes that "an important way to begin learning more about sound environments is to listen more closely to sounds and their relationship to the larger sound world, becoming more aware of changes and imbalances in the acoustic environment." So we have re-activation, re-sensitization, and listening for balance: all very ecologically inspired concepts. While acoustic ecology doesn't explicitly contrast vision and listening, the treatment of sound in the practice of soundwalking is reminiscent of Sterne's (2003) notion of the "audio-visual litany" which, as he argues, essentializes the core qualities of both hearing and vision. Soundwalking in many foundational writings is imagined to unlock dimensions that visualism, in its presumed mono-dimensionality, cannot. One of the more poetic writers in sound studies, Salome Voegelin (2014, p. 1) writes: "the sonic offers itself to experience not as meaning but as a challenge to the making of sense, and that consequently through listening we can reach a different sense, that is plural and includes what we did not expect to know." Another visionary writing coming from critical geography, AM Kanngieser says this about the concept of sound and its role in redressing the ills of the anthropocene:

> Sound is not just about hearing and responding or communicating. It is about becoming aware of registers that are unfamiliar, inaccessible, and maybe even monstrous; registers that are wholly indifferent to the play of human drama. Sound is not only of the human, it undermines human exceptionalism; everything vibrates on some frequency and is touched by vibration, regardless of how imperceptible to human sensibility this might be.
>
> *(2015, p. 82)*

In sum, listening is a new entry point into experience and epistemology, albeit without a formal method or structure, both in the artistic and in the academic domain. In a keynote talk for ISEA, Westerkamp speaks of "disruptive listening" as the paradoxical state of calm and openness in order to jolt ourselves out of habituation and into "perceptual wakefulness" (2015). The disruptive nature of listening is perhaps one of the best ways to gesture to the power of soundwalking: the fact that, despite the absence of prescriptiveness, and perhaps because it is such a subjective experience, listening is a form of presence that is itself disruptive of the status quo. Active listening functions simultaneously as an embodied state and as a metaphor. The metaphor is of listening as an openness to the exterior world to take input in and be disturbed, in contrast to our natural state of being, which is to operate from our interior truths, states, and interests.

In ethnography, soundwalking has been taken up in much more structured ways. To begin with, walking has a long history as a method in cultural studies, and anthropology. In her book on the history of walking, Rebecca Solnit (2001) refers to walking as a "mode of making the world as well as being in it" (p. 29). The social, political, and cultural dimensions of walking that Solnit traces are echoed throughout writings on soundwalking. In 'Listening to a Sense of Place' Jenni Schine[3] traces the nuanced sonic details of place using reflective soundwalking as an embodied interviewing technique. The gendered aspects of walking that Solnit touches on—namely, the practice of taking up space and exercising a freedom of mobility—resonate in Amanda Gutierrez's (2019) practice on soundwalking as a political and intersectional act resisting patriarchal enclosure.

Both walking, and by extension soundwalking, can thus be understood as naturalized everyday practices with political dimensions. At the juncture of acoustic ecology, urban studies, and community engagement, soundwalking has been taken up, along with other "ear-cleaning" methods in a number of contemporary works. Exploring the urban environment with specific interest groups always tends to incorporate some element of soundwalking, whether it is mapping the city with teenagers (O Keeffe, 2015b) or finding acoustic zones suitable for tinnitus sufferers (Atkinson, 2007). These works feature rich descriptions of the approaches used for soundwalking in order to elicit listening impressions from participants towards mapping and characterizing particular soundscapes. In applied acoustics, a field which has been traditionally quite empirical and quantitative, a number of contributors have begun incorporating and (more notably) standardizing participatory approaches such as soundwalking. In these projects, we see an attempt to not only streamline a method for soundwalking in combination with other note-taking, but also to devise analytical frameworks for the dissection and codifying of soundwalking impressions through questionnaires, sorting exercises, and categorization (Aletta and Kang, 2018). One example is Craig, Moore, and Knox's (2017) urban assessment study, which uses Cain et al.'s (2008) framework for in-situ listening evaluation: a framework that incorporates not only participants' demographics, location, and type of space, but also temporal aspects and activity genre. This "experience-sampling" method is enriched with questions about the emotional state of the listener aiming to offer a comprehensive framework for standardizing urban sonic satisfaction. In a sense, soundwalking here is lost in the empirical nature of the study, however the interest in a listening experience within an everyday urban context resonates with the relative objectives of soundwalking: to become conscious of the unconscious act of listening. A more explicit attempt to operationalize soundwalking is evident in Bahali and Tamer-Bayazit's (2017) study of Gezi Park. In addition to a pre-soundwalk questionnaire where participants log expectations of the upcoming listening experience, the study design incorporates a post-survey reflecting on actual experience, and regular stops during the soundwalk where participants are asked to fill out tables logging and categorizing sounds. In my own work, I have also moved away from the more meditative aspects of soundwalking, and towards a more structured data-gathering, in an attempt to formalize complex listener experiences into city planning guidelines. For instance, in a recent (2019–2020) research project in partnership with an architecture firm,[4] I utilized an end-to-end questionnaire with city consultants, planners, and architects as part of a soundwalk event in Vancouver, Canada. The goal was to focus on acoustic features and urban design and to generate some imaginative solutions to sonic pain points in the city. We did not record the soundwalk or encourage participants to gather media, which is a departure from other applied acoustics projects. In all three projects described above, a recording was made of the "actual" soundscape and analyzed separately to compare a presumed "neutral" perspective with the subjective experiences of listener-participants. Still, the reflexive potential of soundwalking in my study was compromised, I feel, by the attempt to collect, sort, and operationalize sonic solutions, to the (somewhat ironic) end of more *livable* urban design. My participants did make important observations about sound in a culturally contested area, namely that sanitation laws have curbed the lively din and sonic energy of open food markets, street food vendors, etc. These comments were in relation to an area which was formerly a black neighborhood (Hogan's Alley) and subsequently Chinatown. Ethnic businesses today are competing with a gentrifying white hipster culture due to its proximity to Vancouver's downtown core (parts of it are dubbed Little Hollywood). This area is also enclosed on all sides by major transport arteries and rapid transit. To that end, another insightful observation from participants was the vertical plane of sound source distribution: for example, the possibility of creating sunken or amphitheatrical areas as a reprieve from traffic and skytrain noise. Still, while instructive and applied, this usage of soundwalking did no justice to the lived sonic experience of place. I recall one soundwalking participant stopped to chat with a shop vendor about a local social justice initiative and kept us waiting for quite some time. As the soundwalk leader, I felt the "sanctity" of the soundwalking format being compromised. But on second reflection, that exchange was much more "ecological" in the cultural sense of the word than our group excursion as sound tourists in this area. In the end, operationalizing any part of this exercise towards piecemeal practical interventions did no justice to the silent refrain of climate change that should be part of any holistic listening practice. But perhaps soundwalking gives justice to the lived experience of people past, present, and future and by noticing ruptures, it could be an act of building bridges between these temporalities. In that, perhaps the most valuable gift that soundwalking gives us is not even listening, but presence: a presence that pulls and pushes listeners along a time axis while simultaneously being firmly and materially located in place.

Ethical considerations of field recording

This brings us to the ethical dimensions of soundwalking and field recording. What *responsibility* do we carry as soundwalkers and attentive listeners to the environment, to the land, and to its human and non-human inhabitants? I don't mean a representational responsibility through compositional interpretation (Wright, 2017); I mean an ethical responsibility to the environment itself (Robinson, 2020). These questions aren't new but they haven't become part of the "canon" of soundwalking in acoustic ecology. This begs the question about the role of sound art, sound research, and the "soundscape perspective" in general, in the movement to reverse the climate emergency. The environmental crisis is a result of systems we have put in place that require us to overproduce, exploit natural resources, valorize disposability, and manufacture the desire for endless convenience. We then counteract these social failings with a turn inward, with meditation, reflection, and attentive observation: we literally counteract this with mediation *technologies* (Hagood, 2019). Soundwalking, to me, falls squarely in that category in that we soundwalk the results of industrialization and modernization, yet yearn for birds and lapping shores. The observations we can make about imbalance or masking in the city are a systemic effect. We are soundwalking extinction in many forms: the extinction of local communities, the extinction of indigenous peoples in Canada, the extinction of plant, fish, and animal species from the environment. It is imperative that we do not do so with self-congratulation for giving our presence and listening attentiveness. It is imperative we do so with the consideration that we are ourselves complicit in that which we listen to and draw inspiration from, for research or composition. At its worst, soundwalking can be complicit in framing ourselves as innocent bystanders of climate change and other forms of colonial capitalist extinction (Kanngieser, 2015). In a recent piece, Hildegard Westerkamp talks about "sustainable soundwalking" as an ethical way of cultivating a personal listening practice. One way of practicing self-reflexivity, she suggests, is by way of "sonic writing": an attempt to describe soundscapes in such detail so as to invite the reader into the sonic world itself. Westerkamp has also talked about the "ears" of the microphone[5]; how its apolitical and a-cultural approach to capture is nevertheless directed by a recordist with interests, aesthetic sensibilities, and political leanings. As a field recordist she reflects on the different stances of her microphone practice: from stationary to moving, to searching. These writings have always struck me as unique in their self-reflexivity. Most recordists, if they talk about microphone position, do so with technical and fidelity-driven considerations only.[6] A notable exception here is Mark Peter Wright's (2017) concept of "post-natural" sound art as a practice that "aims for another type of ecology, one that consistently examines heard and unheard aspects of ethics, power and agency within the practice itself." Among considerations of post-colonial and more-than human relations, Wright creates a new vocabulary for considering the human–technological assemblage of the field recording/recordist coming into contact with what we call nature. He calls them "ethico-political cartographies" in recognition of their place in technological histories and practices of use. So how do we conceptualize a practice of soundwalking that may or may not include audio capture, from a climate-aware, and more-than-human perspective?

A handy concept comes to our aid: Mimi Sheller's concept of "mobility justice," discussed by Scott (2020), which frames justice—social and environmental—not as a static state of affairs but as "a process of emergent relationships in which the interplay of diverse (im)mobilities forms a foundational part" (p. 20). In an article advocating for inter-species mobilities justice, Scott argues that we need a paradigm that not only tackles climate change but also considers "the runaway urbanization and (im) mobilities … flowing from climate change" (Scott, 2020). Mobility here is meant not in opposition to ideas of place and situatedness but as a metaphor for the fluidity of socio-cultural and political structures. Starting from the premise that justice is mobile and involves shifting relationships, Scott proposes three concepts for "citizenship" which could be afforded to non-human species as a way of redressing anthropocene enclosures on various habitats. This would afford e.g. crows and whales the freedom *and right* to move, gather, forage, and dwell in their ancestral areas. While ecology isn't a central tenet of this framework it is woven deeply into the idea of mobilities justice. In that way it connects with listening because, when we listen, we observe the sonic consequences of (failing) social, political, and natural ecosystems. Soundwalking then, is a privileged form of mobility. In order to put that privilege to good use we need to open our ears (and our whole bodies) to more than human worlds and ask ourselves, beyond bearing witness to environments, how do we *do* justice by listening? How can soundwalking be a practice for social and ecological justice? In this chapter I suggest some ways of examining our own relationship to listening, walking, and positionality (Robinson, 2020) contained in three accounts of soundwalking in distinctly different ecosystems. In these vignettes of sonic fiction, I reflect not only on what I hear but also on how I form relationships to audio recording and earwitnessing in the process, and how my listening positionality has shifted historically. I hope to demonstrate a practice of noting temporal ruptures, and emplaced observations that come alongside listening and walking.

Soundwalking extinction series

In 2020, when the pandemic lockdown came into effect, I was unable to go on the one trip I had really been looking forward to: a four-day kayaking tour around the Broken Islands, off the coast of Ucluelet, BC (Canada). Postponing the trip for a year made me look forward even more to the serenity and reprieve of this pristine natural habitat. Then, just before I went, I organized an open conversation between sonic researchers affiliated with SFU (Simon Fraser University) and Dr. Dylan Robinson, broadly about his book *Hungry Listening*, and some directions towards ethical engagements with sound. In the course of the talk, we touched on the use of anthropological audio recording to "preserve" native American (and Canadian First Nations) song and languages: recordings, which Jonathan Sterne calls "resonant tombs" (2003). The idea that these cultural artefacts were to be preserved only accelerated the decimation of indigenous heritage across North America by declaring it dead already, and thus belonging in a museum. On this topic, Dylan Robinson remarked, "well, what does it mean to record natural soundscapes in the age of climate change?" This question instantly struck me as so very relevant for the field of acoustic ecology and my practice of field recording in particular. On one hand, I've always felt extensive audio recording in the field objectifies individual sound sources at the expense of a holistic view of the soundscape; on the other hand, audio recording is a habitual way of collecting little bits of sonic impressions in my soundwalking practice. In the end, after this comment rang in my mind all the way to the Broken Islands, I found myself with a different relationship to sound. Of course, as Mark Peter Wright (2017) notes, Hildegard Westerkamp has been engaging with the ethical dimensions of the *aesthetic* of sound recording for decades. In her own words, "we have to ask ourselves … whether we, in fact, bring our listeners closer to a place or situation or whether we are fooling ourselves and are inadvertently assisting in the place's extinction" (Westerkamp, 1998, p. 75). Wright then issues his own warning regarding the ontology of aesthetic representation with recording technology—specifically transduction devices aiming to amplify unhearable processes: river beds, tree bark, insect bodies, the Arctic. He beautifully articulates the discomfort I have been feeling about these "fashionable" forms of field recording with their gritty sonic materiality reaching for the electroacoustic sublime: "I am willing to accept sonic materiality but not at the expense of political selfhood" (Wright, 2017, p. 5). Often omitted from these transduction recordings is the walking component—the listening body, the movement in space, the "who" of recording decisions, the impact of the sound-walker on their subjects, including the land.

Soundwalking extinction: Broken Islands

I've just been dropped off on a small beach with my kayaking group, and we are about to begin a four-day journey after we pack all our gear in the boats. What I thought I was going to be doing on this trip was relaxing into the serenity and remoteness of the place. The Broken Islands are part of an area stewarded by the Tseshaht nation and only accessible by boat. Their remote location features an almost unique weather front. Even at the height of summer a fog hangs low over the water at least half the day. I thought I was going to be recording little snippets of sound and treasuring them later like gems from a journey to a faraway place. Instead, I find myself in awe of this place, and in a way don't feel I should be here. At every level I am struck by how artificial our presence is, how invasive we are to the dramatic (to my ears) unfolding of everyday life here in the Broken Islands: the life of ravens and seals, sea otters, eagles, oystercatchers, and guillemot pigeons. The eerie fog that we paddle through is quiet and muffles both our visual and sonic fields. Our kayaks feel as if they are slicing through space, our voices echoing across the water and being swallowed by fog. We stop at another part of the island and take a short trip into the forest to a giant old tree. We hear a story from our guides about the Tseshaht who lived here and the ones who lived before them, evidenced by vast shell midden shores and fishing corals. We learn about some of the vegetation around; the structures and villages that used to be. We hear about the lichen on the tree branches: the one that means the air is very clean (Figure 5.1). The environment feels to me like a pristine ecosystem in which we are intruding. The tree has seen and heard a lot. Someone asks if the tree misses hearing people's voices around, children playing, and human life. What an interesting thought: that the tree might be lonely with just its forest inhabitants for company. We take our shoes off and walk barefoot back to the shore. As the sound of my footsteps is absorbed by the soft springy ground, I wonder if our skin is leaving bacteria harmful to the small and delicate mosses of the forest floor. The ocean waves are slowly coating the shore and the tide is rising. It is really only possible to listen on the shore; the wind is too strong while paddling in the open waters. Yet, again, I find nothing in particular to record. Nature isn't spectacular and flamboyant here in the North Pacific; it is understated, blended in, intertwined with the land, the rocks, and the water. It feels as if recording would be just another form of resource extraction. Instead, I appreciate a visit from the Tseshaht beach keepers: they stay for a while to share some traditional Nuu-chah-nulth knowledge. They sing some songs and tell some stories, taking their time between sections. These unexpected silences greatly distress our one 18-year-old group member, who later remarks that he eventually figured out pauses are good, and to be enjoyed—a way of slowing down. The beach keepers share some of their traditional Wakishan dialect, consulting a mobile language revitalization app. A Caucasian man (not part of our group) immediately perks up to ask how to download the app; I cringe. While this area is remote, it is teeming with kayak groups and hikers. I'm very much aware of the anthropocenic impact we are collectively having on the environment, and I don't want to contribute more by extractivist media practices. Instead, I feel drawn to the life unfolding around us: eagles circling above for their evening hunt; seals probably avoiding our shore for their late afternoon sunbathing. My last listening observation as we are picked up by motorboat back to Ucluelet is the vast acoustic profile of the boat's engine. I wonder what the local species think about this intrusion echoing between the shores and vibrating the ocean water below. Our excursion is very environmentally conscious; we leave no trace behind; and yet, I feel grief over our collective footprint on nature, rather than joy at escaping the city.

FIGURE 5.1 Broken Islands 2021

Soundwalking extinction: Ironworks

There is a mysterious periodic buzzing sound in my house (located in a trendy developing area). Agitated, I go around the house unplugging appliances to find the culprit until I finally realize it is a drill or chainsaw coming from one of the many nearby construction zones; it is so faint that it becomes omnidirectional inside. Such is the leitmotif of soundwalking around my neighborhood: insidious mysterious buzzing. Up-and-coming is what real estate agents call my area. That means constant tearing down and reconstruction of older houses, updating infrastructure, and increasing density, which means more traffic and more traffic jams. One of the largest presences on my block is the Ironworks building (Figure 5.2). This revamped industrial business location is a state-of-the-art modern monolith complete with underground industrial storage, private truck entry, and courtyard parking. A thin strip of grass surrounds the building as if to humanize its steely vibe. One of Schafer's (1977) most insightful observations in *The Tuning of the World* was the omnipresence of electrical hum as the new "tonal centre" of modernity. Beside their mechanistic uniformity which sets them apart from organic sound sources, electrical hums coming from the Ironworks building are anything but identical. There are several different frequency bands of buzzing; an almost whistling complex tone with (I suspect) triangular waveform; another with a rough, ringing timbre; a soft electrical hum from the wall of hydro power meters; various clicks and beeps from automated doors, hinges, and hatches; reverse beep sounds from trucks and other service vehicles, and, on any given day, other unique machine drones. Through iteration of soundwalking—that is, by walking and listening again and again and trying to be attentive to my positionality, and different histories present here—I realize there is a significance to the different hums and drones. The ones coming from the Ironworks building are uniform and constant, wired into the high-tech industrial construction of the building. But there are other, smaller creaks, buzzes, and intermittent drones coming from the multitude of small family-owned shops: storage facilities, auto body shops, metallurgy. Those sounds are almost organic: the old buildings have creaky metal doors and mechanical springs that produce a very different industrial soundtrack.

FIGURE 5.2 Pigeon friends and the Ironworks building corner

Soundwalking in this area has always inspired me to meticulously record and document my experience in an effort to understand and build an informal dataset of changes in the soundscape signaling economic and cultural changes in my community. Ironworks is located near the Port of Vancouver: a central depot for cargo ships and trains, delivering fish, sugar, fuel, and grain. The area, as a result, maintains a large contingent of seagull, crow, and pigeon populations; an avian community that characteristically adds to the soundtrack of this area. The juxtaposition between the drama (as I hear it now) of bird calls and the multitude of artificial mechanistic sounds, plus all-day traffic from a nearby main road, invites my ethnographic ears. I feel responsible to understand and intervene in this habitat somehow. I feel ownership of this neighborhood and am deeply embedded in its sonic life, human and non-human alike. In the beginning of the pandemic, because I so regularly do audio recordings around Ironworks, I was able to capture and represent, using audio as well as spectrographs, the hushing of din and traffic as a result of the Spring 2020 lockdowns. Like many acoustic ecologists have noted,[7] the shift in urban soundscapes was audible and palpable; perhaps not only because of the cessation of work and commerce, but because of the shift in consciousness, the turning inward, the collectivisation of anxiety and slowdown. Like many people, I soundwalked regularly through my environment. At first, the Covid-19 lockdown felt like soundwalking the extinction of industrial sounds. The former vibrancy of my mixed-zoning neighborhood was replaced by quietude and birdsong. In my most recent soundwalk, however, I am finding a different kind of extinction. Alongside the vibrant Ironworks building there are the faint idiosyncratic sounds of small ethnic businesses: food storage, supplies, auto body shops, metalwork, deliveries. These buildings are old and creaky, groaning under the weight of their years of rain and decay. Rather than electronic drones, the sounds coming from these businesses are just as industrial yet somehow alive. I realize as I listen that these old soundmarks of traditional small commerce are becoming extinct, overshadowed by the vast buzzing energy of the Ironworks mega-complex. The sonic extinction signals real economic distinction. As this community develops further, these businesses and their sounds will continue to be displaced, the traditions, relationships, and local flavor replaced by the corporate energy of the Ironworks clientele or masked by the racket of unending destruction and construction of more condos and trendy microbreweries in their place. This smaller subtle extinction leaves me wondering: do the birds care? Are local avian species concerned with this economic shift? Does it spell unwelcome change for them too, robbing them of shelter and ancestral nesting areas, from which to crow their avian thoughts?

Soundwalking thriving worlds: rock doves in Varna

In October 2021 I find myself on a research trip in Bulgaria, my homeland; the pandemic shows no signs of slowing down, yet the world has moved significantly since the first lockdown. Despite the global collectivisation of pandemic measures, each country and each individual has experienced their own pandemic challenges and adaptations, within their larger community. In Bulgaria, life is almost back "to normal" after a summer of hardly any Covid-19 restruction, so as I soundwalk around my hometown of Varna I expect a "normal" soundscape. My experience gives me pause, however. More-than human and indigenous perspectives of sound have been on my mind, and I am listening differently here and now. I'm also visiting outside the festive summer hours that I usually come here for. Life is more subdued, more routine; the music streaming out of cafes and restaurants is not as loud; the murmurs of crowds much thinner. Something I've noted on my visits before as the main contrast to my North American sonic experience is the absence of HVAC (heating, ventilation, and air conditioning) sounds here. In Vancouver, soft narrowband noise streams from every vent on every building, creating a blanket of pink noise and enclosing the acoustic horizon of the listener. For this reason I've always loved coming to Bulgaria and feeling my body relax with the absence of these vibrations all around me. One of the big changes in the last decade in Bulgaria, however, is the installation of individual AC units on a mass scale. Building standards here have never included proper insulation, heating, and cooling, and people's living standards have finally allowed for this "Western" perk. I have been dreading the broadband noise pollution to come as a result, but I am quickly reminded that electricity is too expensive. These units are run for minutes at a time and then stopped. My listening horizon remains open. What these outdoor units provide, however, is additional perching and shelter for the city's birds. The birds of Varna have been subject to much interest and attention.[8] Proximity to the seashore and the existence of a large botanical garden in the heart of the city is responsible for healthy bird populations. Seagulls are most certainly a soundmark of Varna, and with the colder weather they have moved off the shore and into the city for food scraps. They make a boisterous racket in every street square, showing each other up. The baby birds with grey speckled coats are the loudest, shrieking in a higher pitch than their parents. Street cats are quiet, observing from a distance. They might meow for food but each urban scavenger knows their turf. Still, what I am listening to most attentively is pigeons. Pigeons have become my Covid lockdown obsession ever since a pigeon self-rescued on my porch in May 2020. Over that year I got to know the sound of Dovey's various coos and vocalizations, the flap of his wings, his moods, and molts (Figure 5.2). We all went a little bit mad in lockdown, didn't we, but also more observant and caring? It was a journey getting to know Dovey's story. He aggressively pursued my company and I eventually found out he was my neighbor's rescue. Every time he cooed differently I was on the Internet researching pigeon vocalization. When Dovey brought some friends over to my balcony I could actually distinguish the sound of his particular wing flap at takeoff compared with those of his bird pals. I learned common city pigeons are called rock doves and are essentially domesticated birds with a highly evolved social and community life. I am amazed now, walking in the city of Varna, that I hear the nesting coos of city doves for the first time. These are not the sharper, more intoned sounds of the mourning dove I've heard before, but soft guttural, rhythmic coos meant to attract a mate to the nest. These coos have always been here, yet I've never heard them, because they never registered on my anthropocenic existence. Simply put, I never looked up to listen.

Pigeons like to roost in high places, and the multitude of aging buildings in Varna provide a perfect habitat. In contrast with the German minimalism of Vancouver's high-rises, many three-storey buildings in Varna feature decorative roofs and balconies, with lots of nooks and crannies, hosting families of cooing pigeons. As I approach the train station (Figure 5.3)—one of the most meaningful locations of my childhood since trains led to everything from grandparents to summer camps—my listener training anticipates the reverberant acoustics of the old majestic dome structure. I have recorded here before on my previous visits: the long screech of train wheels; the rhythmic pace of moving trains; the intense echo and resonance of the train announcements inside the train station. Distorted sound that every traveller is painfully used to, so loud and so affected by echo that it is almost impossible to make out the actual announcement: yet the voice is nostalgically familiar as an ageless recording I remember from my childhood. Amidst the hustle and bustle of travellers and distorted PA announcements, I now hear a symphony of reverberating wing flaps, cooing, and courtship calls. This is pigeon central: the many covered ledges are homes to nests where parents are feeding their young. Even inside the building, pigeons roost high up into the dome ceiling, their coos amplified by the acoustics. Inside, people come and go, buying tickets, water, snacks. I have done so myself countless times but as I soundwalk now I am discovering these sounds for the first time. I wonder if the regular employees here are more aware of their non-human companions; if they welcome them or resent them as pests. As every pigeon enthusiast on Instagram will tell you (yes, this subculture extends to social media), pigeons were domesticated during various wartimes as courier birds, then released to fend for themselves. Hearing them through this new lens, I think, they have just as much right to be here as we do. The pigeon community in Varna is like a parallel universe, thriving and unconcerned with human affairs, yet very much dependent on human structures for shelter and food sources.

FIGURE 5.3 Varna train station and pigeon nests in the dome ceilings

Conclusion

To me, Dylan Robinson's remark about recording nature in the age of climate extinction has the same implication that recording native song had for North American indigenous communities: the act of recording itself is a sentence to extinction. The impetus to record and raise awareness has underlying motivations driven often by either aestheticization or the desire to preserve. As Mark Peter

Wright puts it (2017), we need to turn the microphone inward and dredge up "a history of phantasmagoric subjectivities and media captures in order to address urgent ecological matters" (p. 5). To be reflexive of the impulse to record sound means to realize we are listening on borrowed time. For that matter, we soundwalk on borrowed time. Yet, soundwalking has the potential to be a powerful political act of (ear)witnessing and mobilities justice. Two necessary conditions for that, I propose, are listening positionality, and critical self-reflection: a kind of listening to how we listen. Neither idea is exactly new; it has featured in many soundwalking practitioners' writings, particularly those who already occupy more marginalized positions (McCartney, 2016; Brownell, Sheridan, and Scales, 2018). The question is, how to square that with the "mainstream" practice of soundwalking within acoustic ecology, which has always involved a measure of mysticism and aesthetic appreciation of nature's sounds. There is a valorization of both ecological clarity in pristine environments (Monacchi, 2014; Krause, 2012), and a compositional interest in the complexity of organic sound (Truax, 2015). How to heed Wright's warning against "sonic selfies" (2017, p. 5)—a centering of ourselves and our creative vision in representing soundscapes? When Westerkamp speaks of sustainable soundwalking (2018) and disruptive listening, the question is how and to what end, given the pressing issues we are faced with: decolonization on one side, and climate extinction on the other. If I am to synthesize the common threads of my three case studies, beyond principles of positionality and reflexivity, I would add a critical temporal awareness as well. In each case it is *because* I listened reflexively that I noticed sounds I hadn't noticed before (e.g. the pigeons in Varna)—and that allowed me to consider what made me tune them out before? Thinking about issues of decolonization and climate change caused me to have a very different relationship to remote nature in the Broken Islands, which was a springboard for questioning my role as a tourist and my physical impact on the environment. Finally, listening to my industrial neighborhood led me to thinking through issues of past and future urban developments and gentrification, and what community resilience means in this context. Soundwalking, as I suggest at the beginning of this chapter, is a bridge between these different temporalities but only if the engagement is authentic, if the ears are open to questioning our most foundational listening assumptions.

Soundwalking, for me, has always been an ethnographic activity. As an immigrant and a queer person—in other words as someone who doesn't fit in the mainstream already, soundwalking has always been a reflective activity and a knowledge gathering exercise that connects me to place and to networks of beings. It is an activity that is removed from heteronormativity enough that it feels simultaneously queer, safe, and new: it allows me to be myself in an uncomplicated way so that I can complicate my presence and my relationships to my human and non-human surroundings. In addition to soundwalking in the "real environment" I soundwalk in another, rather unconventional, context: game worlds. Games provide an interesting platform for speculative sonic fiction with the spaces they create. In fantasy role playing games, soundscapes are extremely complex, mimicking the layering and randomness of the "real world" and as a result the experience of virtual soundwalking feels both scripted but also rather authentic. I have, for years, shown examples of soundwalking in game environments when I teach acoustic ecology. I ask students to reflect on the intended listening positionality of the player-avatar; to identify what is present and what is missing in the environment sonically; as well as what sort of place is being constructed and scripted with sounds. Knowing this interest in games, Hildegard Westerkamp recently brought a new independent game to my attention, titled *Seasons.*[9] *Seasons* is not a game about soundwalking, not exactly. In the demo notes (the game is still in development), climate change is quite advanced, and the extinction of habitats and seasons is somewhat inevitable. The player navigates through environments on a bike and the main game action is capturing—within the game—a media library of last impressions from nature. Using photography and field recording, the character is a *witness* to profound change rather than a savior of the world. It is yet unclear to me whether the game has pre-determined objects that will need to be "collected" or function more as an open world for a media historian. What fascinates me, however, is the possibility offered by atmospheric game mechanics to convey a nuanced emotional sensescape and challenge the player with novel mechanics such as listening and recording. The very fact of doing this in a technological medium is of course paradoxical from an environmental sustainability standpoint: a game about seeing and listening foregrounds the inherent paradox that we can't truly give justice to that which is going extinct. The game concept returns me to the ethical dilemma stated earlier: what does it mean to record natural sound in a time of climate extinction? What does it mean to capture the end of the world in a game given the sort of consumerist media ecology that games are a part of? Would playing *Seasons* be a form of escapism or a willful attempt to deal with climate grief for a generation too powerless to intervene in concrete ways? As an ethnographic listener, my soundwalking is often already a kind of game: hunting for clues and collecting objects, basking in the emotional space that listening affords. The time has come, however, for me to stop and reflect on my own practice.

In the three soundwalking vignettes above, I employ what Westerkamp calls "sonic writing" (2018) to draw the reader into my listening and reflect on how my listening positionality has changed over time, and I demonstrate how that draws out different critical reflections on history, culture, economy, etc. These shifts have led me to not only "discover" different sounds but also to question my relationship to technology and recording. I trace how concerns over climate change, and the ethics of presence in space, can alter one's listening positionality in a way that reveals "hidden registers" of existence (Kanngieser, 2015). I hope to show that by critically reflecting on what we hear, and what we didn't hear before, we could hope to perform a type of mobilities justice rooted in temporal ruptures. These listening observations must lead to action of course: at the very least to minimize our own impact on the environment and to reflect on the delicate ecological networks of which we are just one part, no bigger and no smaller than any other. By fully assuming responsibility for our role in climate change and injustice, and by shedding our anthropo-centrism, soundwalking can serve as a reckoning, a witnessing, and a formulation of individual, then collective, responsibility to deal with current issues. As an emplaced act of reaching towards tensions and temporalities it can become an exercise in consciously de-centering ourselves; then it can be a practice for ecological stewardship, towards effecting change—otherwise we are only soundwalking extinction.

Notes

1 See the work of Leah Barclay for instance, for use of embedded mics, hydrophones, and mobile applications: http://leahbarclay.com/portfolio/
2 See http://facs-newmedia.finearts.yorku.ca/andra/soundwalk/qep.html
3 See http://artaban.ca/work/listening-to-a-sense-of-place
4 See https://www.sfu.ca/sonic-studio/livable-soundscapes.html
5 See https://www.hildegardwesterkamp.ca/writings/writingsby/?post_id=74&title=the-microphone-ear:-field-recording-the-soundscape
6 This is true in even very socially and politically conscious communities such as Cities and Memory: https://citiesandmemory.com/field-recording-guides/
7 See https://theconversation.com/how-covid-19-shutdowns-are-allowing-us-to-hear-more-of-nature-136139 and https://www.nytimes.com/interactive/2020/05/22/upshot/coronavirus-quiet-city-noise.html?smid=tw-share
8 See https://www.moreto.net/novini.php?n=253699 (in Bulgarian)
9 See https://blog.playstation.com/2020/12/10/introducing-season-a-game-about-capturing-that-fleeting-moment/

References

Aletta, F. and Kang, J. (2018) Towards an Urban Vibrancy Model: A Soundscape Approach. *IJERPH* 15, 1712. doi:10.3390/ijerph15081712

Atkinson, R. (2007) Ecology of Sound: The Sonic Order of Urban Space. *Urban Studies* 44, 1905–1917. doi:10.1080/00420980701471901

Bahali, S. and Tamer-Bayazit, N. (2017) Soundscape Research on the Gezi Park—Tunel Square Route. *Applied Acoustics* 116, 260–270. doi:10.1016/j.apacoust.2016.10.002

Biserna, E. (2015) Mediated Listening Paths: Breaking the Auditory Bubble. *Wii Journal of Mobile Media* 9, 30.

Brownell, C.J., Sheridan, D.M., and Scales, C.A. (2018) (Re)Mediating (Un)Heard Resonances: Tracing the Rhythms of Aurality in a Residential College Community. *Educational Studies* 54, 396–414. doi:10.1080/00131946.2018.1476355

Cain, R., Jennings, P., Adams, M., Bruce, N., Carlyle, A., Cusack, P., Davies, W., Hume, K., and Plack, C.J. (2008) SOUND-SCAPE: A Framework for Characterising Positive Urban Soundscapes. *The Journal of the Acoustical Society of America* 123(5), 3394–3394. doi: 10.1121/1.2934071

Ceraso, S. (2014) (Re) Educating the Senses: Multimodal Listening, Bodily Learning, and the Composition of Sonic Experiences. *College English* 77, 102.

Craig, A., Moore, D., and Knox, D. (2017) Experience Sampling: Assessing Urban Soundscapes Using In-situ Participatory Methods. *Applied Acoustics* 117, 227–235. doi: 10.1016/j.apacoust.2016.05.026

Droumeva, M. (2017) Soundmapping as Critical Cartography: Engaging Publics in Listening to the Environment. *Communication and the Public* 2(4). doi:10.1177/2057047317719469

Droumeva, M. and Jordan, R. (2019) *Sound, Media, Ecology*. New York, NY: Palgrave Macmillan.

Gutierrez, A. (2019) Flâneuse>La caminanta. Sounding Out! Available at https://soundstudiesblog.com/2019/08/12/flaneusela-caminanta/ (accessed November 25, 2021).

Hagood, M. (2019) *Hush: Media and Sonic Self-Control*. Durham, NC: Duke University Press.

Haraway, D. (2016) *Staying with the Trouble*. Durham, NC: Duke University Press.

Kanngieser, A.M. (2015) Geopolitics and the Anthropocene: Five Propositions for Sound. *GeoHumanities* 1, 80–85. doi:10.1080/2373566X.2015.1075360

Krause, B. (2012) *The Great Animal Orchestra: Finding the Origins of Music in the World's Wild Places*. New York, NY: Little, Brown and Company.

McCartney, A. (2016) Ethical Questions about Working with Soundscapes. *Organised Sound* 21, 160–165. doi:10.1017/S135577181600008X

Monacchi, D. (2014) Fragments of Extinction—An Eco-acoustic Music Project on Primary Rainforest Biodiversity. *Meedizioni* 128.

O Keeffe, L. (2015a) Listening to Ecological Interference. Dark Mountain. Available at: https://dark-mountain.net/listening-to-ecological-interference-renewable-technologies-and-their-soundscapes/ (accessed April 27, 2022).

O Keeffe, L. (2015b) Thinking through New Methodologies. Sounding Out the City with Teenagers. *Qualitative Sociology Review* 11(1), 6–32.

Robinson, D. (2020) *Hungry Listening: Resonant Theory for Indigenous Sound Studies*. University of Minnesota Press.

Sterne, J. (2003) *The Audible Past: Cultural Origins of Sound Reproduction*, Durham, NC: Duke University Press.

Schafer, R.M. (1977) *The Tuning of the World*. New York: Knopf. Reprinted as *Our Sonic Environment and the Soundscape: The Tuning of the World*. Destiny Books, 1994.

Scott, N. (2020) A Political Theory of Interspecies Mobility Justice. *Mobilities* 15(6), 880–895. doi:10.1080/17450101.2020.1819728

Solnit, R. (2001) *Wanderlust: A History of Walking*. London: Penguin Books.

Taylor, S., and Fernström, M. (2017) Acouscenic Listening and Creative Soundwalks: Evoking Memory and Narratives through Soundscape Exploration. *Leonardo Music Journal* 27, 3–6. doi:10.1162/LMJ_a_00999

Truax, B. (2015) Paradigm Shifts and Electroacoustic Music: Some Personal Reflections. *Organised Sound* 20(1), 105–110.

Voegelin, S. (2014) Soundwalking the Museum: A Sonic Journey through the Visual Display. In: N. Levent and A. Pascual-Leone, eds. *The Multisensory Museum Cross-Disciplinary Perspectives on Touch, Sound, Smell, Memory, and Space*. New York, NY: Rowman & Littlefield.

Westerkamp, H. (1998) Speaking from Inside the Soundscape. Proceedings of the *Hör Upp! Stockholm Hey Listen! Conference on Acoustic Ecology*, June 8–13, 1998.

Westerkamp, H. (2015) The Disruptive Nature of Listening. Keynote Address at the *International Symposium on Electronic Art*, Vancouver, BC. Canada, August 18, 2015.

Westerkamp, H. (2018) Sustainable Soundwalking—Passing On and Relaying Acoustic Ecology's Core Practice. Proceedings of *The Global Composition 2018*, Conference on Sound, Ecology, and Media Culture, Media Campus Dieburg, Hochschule Darmstadt, Germany, October 4–7, 2018.

Westerkamp, H. (2001/1974) Soundwalking. Originally published in *Sound Heritage*, Volume III Number 4.

Wright, M.P. (2017) Post-Natural Sound Arts. *Journal of Sonic Studies*. Available at: https://www.researchcatalogue.net/view/292319/292320/0/0 (accessed April 27, 2022).

6

AURAL BORDER THINKING AS A DECOLONIAL SOUNDWALKING METHODOLOGY

Amanda Gutiérrez

Introduction

I am interested in the soundwalk as a methodology considering the epistemologies, cultural subjectivities, and walking formations from indigenous and migrant experiences in everyday spatial practices. In creating these methods, I will consider narrating personal oral histories in the context of listening positionality (Robinson, 2020) in site-specific projects. This exercise aims to highlight an environment's political and social dimension by enacting situated listening in space to define aural relations with the self. This methodology is primarily informed by listening positionality, which embraces storytelling to recall aural subjectivity. My argument is that listening positionality can be adapted to interdisciplinary and intersectional practices hand in hand with the critical theory of Gloria Anzaldúa. Her writing opens the opportunity for me to conceive an *aural border thinking* methodology in soundwalking, which considers indigenous, migratory, and settler experiences of coloniality that we carry in our ancestry. The development of the concept that I propose is a long-term project informed by walking self-reflexivity (Sotelo, 2010), feminist intersectionality (Crenshaw et al., 2013), critical listening positionality (Robinson, 2020), and border thinking (Mignolo, 2012). Embracing listening positionality as a methodology in soundwalking practice requires enacting walks in groups. This formation employs collective listening and storytelling, dissecting our different positionalities through self-reflexivity in a group walk. As Dylan Robinson has explained: "Developing an awareness of listening positionality here holds potential for listening otherwise, yet the question remains of how—or the extent to which—we might orchestrate such stratified positional listening toward intersectional antiracist, decolonial, queer, and femininist listening practices" (2020, p. 60). In response to the answer of how and which, methodologies are meaningful sources to open a dialogue as researchers, artists, and facilitators for spaces to listen.

DOI: 10.4324/9781003193135-7

To frame decolonial aural thinking, we need to unpack the theoretical background of the concept. Decoloniality comes from specific anti-colonial roots:

> Whereas postcolonial thinking tends to focus on the impact of European colonial rule around the world in the eighteenth through the twentieth centuries, decolonial thinking, particularly the body of thinking that emerges from Latin American and U.S. Latinx contexts, appeals to 1492 as a point of departure.
>
> *(Velez and Tuana, 2020, p. 6)*

Decoloniality seeks to imagine and conceive a future where *otros mundos son posibles*/other worlds are possible outside the colonial hegemonic epistemology (ibid.). In the exercise of socially engaged projects such as the soundwalk, it is eminent to consider genealogies and epistemologies from marginalized experiences oppressed by coloniality. These emancipatory practices must be enacted by a theory that breaks with knowledge extractionism and settler colonialism, named as "hungry listening" (Robinson, 2020).

In creating these epistemologies, it is eminent to consider decolonial feminist methodologies, such as the work of Argentinian philosopher Maria Lugones. She proposes an infra-political way, which constitutes the power of oppressed communities to imagine the construction of their meaning (Lugones, 2010). She suggests moving methodologically from women of color feminism to decolonial feminism and from the grassroots to the "colonial difference" (ibid.). This shift can happen by emphasizing the historical relations of intersubjectivity and its relation to the grounded social movements that confront, critique, and resist the racialized, colonial, and capitalist heterosexual gender oppressions. Lugones focuses on the relation of the subjective-intersubjective to visualize how colonized agencies are disaggregated. Therefore, decolonial feminism is a critical response "where coloniality of power is enacted" (ibid.) while asserting geopolitics of knowledge to build commonality. This commonality opens a space where local histories invent and implement designs, entangling the local with the global, through dialectic situations where the "border thinking" works (ibid.). As Walter Mignolo points out, there is a need to name a critical concept which reflects their own formations:

> For precisely this reason, border thinking that leads to decoloniality is of the essence to unveil that the system of knowledge, beliefs, expectations, dreams, and fantasies upon which the modern/colonial world was built is showing, and will continue to show, its unviability.
>
> *(2012, p. ix)*

Border thinking is not situated in the conception of territory, national borders, or academic disciplines but as a critical dimension of coloniality. Border thinking considers a turn of epistemologies embracing incarnated experiences of the "colonial difference" (Icaza, 2017). The concept opens the possibility of engaging with epistemological, ethical, and aesthetical political forms, liberated and conceived outside hegemonic epistemes of universalism (Mignolo, 2012).

This chapter works as a first attempt to name a decolonial methodology in practice. The sections delve into creative contexts conversing with aurality and the concept of the borderland. The goal of aural border thinking should not be limited to empathy and theory but also to embrace soundwalking practices as political actions. In its construction, I practice embodied methodologies developed through aural pedagogies to open engagement with participants. In this chapter, I reflect on my initial practices while leaving space for its development.

Ancestry through situated listening

To start the theoretical and practical relationship of soundwalking, I will reflect upon the life of the first woman who spoke to me about the importance of walking in her everyday life. My grandmother Leonila remembered her mother's extraneous work of selling bread weekly. Leonila told me about these walking journeys during dinner or when we visited the cemetery to bring flowers to her mother's grave. Her name was Gregoria Lopez, alias Goyita, born in 1908 in the town of Calpulalpan de Méndez in the Zapotec region of Oaxaca. She was a peasant and baker who had to walk six hours daily to reach the town of San Pablo Guelatao to sell bread to provide an income for her family. My grandmother's stories about walking filled my head with questions and possible scenarios of the meaning of walking alone as a woman in the woods. The mountain region where they lived is called the Sierra de Juarez. The area is part of the Sierra Madre of Oaxaca, which holds one of the vastest biodiversity flora and fauna ecosystems, including subtropical evergreen forest, pine-oak, the ocelot, jaguar, brocket deer, and other endangered species. Culturally speaking, the Zapotecs of the North call themselves bene xon, divided into four subcultures: Cajonos, El Rincón, Ixtlán, and Chapman. My grandmother's family came from the Ixtlán municipality, preserving most of their subtropical ecosystem and Zapotec language.

My grandmother told me how Goyita Lopez, my great-grandmother, had an excellent and well-tuned ear, which helped her understand the Ixtlán mountain's clues and sound ecologies. For example, she knew when it was dusk by hearing the owl. Goyita said that the birds' singing whispered along the mountain's walking path. I remember her evoking the different guttural bird calls from the Sierra Juarez, emphasizing the night dove as an essential clue to return home. Or the memory of how she could hear her steps on the icy road during winter, listening and orienting herself inside the vast forest. For Goyita, walking was a tool of mobility and survival. Her walk depended on forms of living communication within the Zapotec Sierra region. These forms of communication were transformed when she moved to the urban roar of Mexico City's streets. Migrating from the subtropical forest of Ixtlán to Mexico City was emotionally and bodily traumatic. Once she immigrated to the urban landscape, she rarely heard the

Zapotec language and the native birds from Ixtlán again. Her experience made me think about spiritual and sacred practices which had to be adapted and, most of the time, lost for First Nations individuals who immigrate to urban environments.

For Goyita, walking and learning new cues in the city were an adaptation process that took decades to attune. This knowledge was passed to me through oral history and everyday walks around the neighborhood of Azcapotzalco in Mexico City, where I grew up as a child. As an adult, I emigrated from Mexico to the United States, settling in Chicago for almost 18 years. I also tuned my ears to new cultural codes and languages on my journey. We both adapted as immigrants to landscapes and cultures that challenged our familiarity with space but provided us with a new set of skills to survive. I am aware of the different experiences that my grandmother and myself have had. Even if I emigrated and struggled as a racialized woman, my social class gave me the privilege of living with a migrant status that allowed me to earn an academic degree. Therefore, my positionality as an immigrant differs from the experience of being undocumented in the United States and Canada. This self-referential positionality (Simandam, 2019) is essential because it brings a space to investigate our lines of history by assessing our privileges and dimensions of oppression. By asserting my own "self-referential positionality," I can identify the constellations of my situated knowledge (Haraway, 1988), which is essential while executing artistic practices in sound ethnographies, especially concerning decolonial methodologies. In terms of the sonic dimension of the self, I propose critical listening positionality as a tool to assert a decolonial practice on sound and music perception, especially while intending to apply intersectional feminism since "engaging in critical listening positionality involves a self-reflexive questioning of how race, class, gender, sexuality, ability, and cultural background intersect and influence the way we are able to hear sound, music, and the world around us" (Robinson, 2020, p. 10). As part of the methodologies that I propose, it is essential to consider a critical listening positionality as a subjective orientation of situated knowledge in the perception and interpretation of sound culture.

While living in Canada, I recognized my settler colonial listener positionality as an immigrant with a particular perception embedded while living in Mexico and the United States. An example of this perceptual phenomenon is my own embodiment experience with the urban soundscape in Mexico City. Their everyday sonic amplitude and density give me a sense of belonging and hype energy. However, this aural hybridity was not audible in a homogenous cultural environment in some cities in the Global North, which in my case creates feelings of auditory alienation. Therefore, reflecting and naming other spatial and cultural experiences of acoustic ecologies opens a decolonial turn, moving away from the aural Western universality associated with wellness in quietness.

Framing Gloria Anzaldúa's literature on positional listening

The relationships of migration among new and adopted environments create forms of knowledge that inform our personal history through walking and listening. Chicana feminist writer Gloria Anzaldúa, in her book, *Borderlands/La Frontera: The New Mestiza* (1987/1999), opens a literary dimension of walking and listening as a poetic experience of being Chicana in the Southern United States. These observations are palpable in her statements and modes of conceiving reality as a nomad. She writes about the experience of Chryscos, a Menominee First Nation and two-spirit poet's sacred listening through walking:

> Four years ago, a red snake crossed my path as I walked through the woods. The direction of its movement, its pace, its colors, the "mood" of the trees and the wind and the snake—they all "spoke" to me, told me things. I look for omens everywhere, everywhere catch glimpses of the patterns and cycles of my life. Stones "speak" to Luisah Teish, a Santera; trees whisper their secrets to Chryscos, a Native American. I remember listening to the voices of the wind as a child and understanding its messages.
>
> *(Anzaldúa, 1987/1999, p. 36)*

Anzaldúa's literature evokes soundscapes from an indigenous and migrant positionality that organizes the sense of hearing as directly connected with the ecology of a particular place. In this case, from the situated memory of Chrycos' experience evoking the sound from an embodied standpoint. Anzaldúa's literary work asserts these embodiments from a situated perspective. Her observations are relevant in constructing a decolonial form of listening and walking that includes the migrant and indigenous subjectivities as powerful allegories of movement. Since the indigenous and the migrant share the unstable experience of preserving and searching for a homeland in a perpetual state of flux, their walks exist in a liminal state of adaptation and sovereignty.

The experiences of my great-grandmother Goyita, like many other experiences of immigrants, are narratives of adaptation linked with sonic memories. Tracing these sonic memories can help us to identify our cultural heritage as a reflective exercise to approach situated listening. Its reflection aims to raise awareness of our political, symbolic, and relational bonds with the indigenous and migrant presence. Listening and thinking collectively about the experience of the itinerant and displaced opens a critical perspective on our cultural and social positionality. In participatory walks, the reflection on the racial, social, and cultural dimensions of settlements allows us to assert self-referential positionality (Simandam, 2019) with the land where we live.[1] Methodologies maintaining an intersectional perspective can help us orient these reflections through positional listening and walking practices. We can conceive this listening tool as an exercise of social consciousness within the environment by hearing while walking, the site-specific soundings as a sensorial and engaging tool (Pink et al., 2010). In this engagement, storytelling narratives place our different positionalities in dialogue by being silent or verbally active (Sotelo, 2010). In Anzaldúa's work, I see the potential to evolve positionality from creative writing and critical race theory to soundwalking practices.

Scholar Gloria Wekker, in her essay, "The Arena of Borders: Gloria Anzaldúa, Intersectionality and Interdisciplinarity" (2018), takes two essential readings from Gloria Anzaldúa's work on the potential of writing as interdisciplinary and intersectional practices. On the one hand, Anzaldúa's framework on feminist intersectionality considers her political and theoretical standpoints regarding gender, ethnicity/race, class, and sexuality. On the other hand, Wekker points to Anzaldúa's situated experience of being lesbian and Chicana as an integral part of her work, which informs cultural traces involving language while writing about her identity as an activist and poet in the Chicano movement. The term *deslenguadas*[2] (Anzaldúa, 1987) refers to the linguistic experience of living between Mexican and North American cultures, resulting in Spanglish as a hybrid language. Deslenguadas is a literary figure that reflects on the experience of cultural adaptation as a first-generation immigrant:

> Deslenguadas. Somos las del español deficiente. We are your linguistic nightmare, your linguistic aberration, your linguistic mestizaje, the subject of your burla. Because we speak with tongues of fire we are culturally crucified. Racially, culturally and linguistically somos huérfanos—we speak an orphan tongue.
>
> *(Anzaldúa, 1987, p. 58)*

By considering language as one of the primary vehicles for storytelling and positional listening, the concept of *deslenguadas* names alternative modes of self-expression, challenging set identities and linguistic oppressions. Making space for cultural hybridity in the borderlands is not only delimited to language but also challenges ideological dichotomies of gender, ethnic, racial, and cultural formations. Anzaldúa's political standpoint on hybridity is reflected in this interview extract:

> I didn't want it to be assimilated so easily, so I code switch, use a lot of Spanish and switch genres—I go from poetry to journalism, to myth, to history, to folk story, to lived experience, to theory. Shifting disrupts the causal meaning.
>
> *(Hernández and Anzaldúa, 1995, p. 13)*

Wekker (2018) bridges this approach in *Borderlands* as an analogy of Anzaldúa's migratory ideas and writing practices that has influenced other disciplines. The fluidity in her interdisciplinarity writing is reflected through critical theory, poetry, and her editorial work, which invited a creative and political cross-pollination among Women of Color writers, painters, and performance artists.[3]

Gloria Anzaldúa's interdisciplinary work can serve as the foundation for a soundwalking methodology that migrates among fields of study, cultural references, and hybrid genres that are present in her book *Borderlands*, where her poetry navigates to critical theory, while history is embedded into autoethnography.

The meaning of aural border thinking that I propose takes into consideration the use of interdisciplinary practices incarnated in the exercise of critical listening positionality and walking. Aural border thinking departs from the subjective sounds of individual experiences on migration journeys. Therefore, the goal of aural border thinking is to open a space for self-referential positionality, delving into the entangled experiences of the migrant, indigenous, and settler.

My goal is to arouse these experiences through walking and aural pedagogies responding to site-specific cultural and social dimensions. Aural pedagogies are an entry point to sensorial and ethnographic exercises in collectivity. These pedagogies bring a critical question of how voice, argumentation, race, and power (McRae and Nainby, 2020) are implicit in our aural culture. Their methodologies expand on radical modes of engagement through sound knowledge and auditory perception. Their execution invites collective and situated listening to perceive aural embodiment as a learning practice. In my research creation, aural pedagogies are a crucial tool that challenges colonial forms of listening and walking as creative outlets. In the next sections, I will describe three projects that approach these methods as a tool of engagement.

Listening/Speaking in Tongues (2022)

I apply the methodology of aural border-thinking in my soundwalk titled: *Listening/Speaking in Tongues* (2022), which departs from Anzaldúa's essay, "Speaking in Tongues, a Letter to Third World Women Writers" (in Moraga and Anzaldúa, 1981/2015). The soundwalk critically relates to my positionality as a Concordia University student in Montreal, finding symbolic connections between experience and struggles in academia. The soundwalk points to political events and meaningful spaces for Communities of Color in Concordia.

The walk started in the library of Concordia. Around 15 people attended. Most of the audience were students and faculty from the university. The group of participants responded to the invitation of the university *Soundwalk Symposium 2022*; therefore, they seemed to know what to expect a soundwalk to be, having previous experiences with the aesthetics of walking and listening. Those who were there for the first time had a silent engagement with reading the Spanglish lines of Anzaldúa. Nevertheless, the walk opened a space to trace the relationship between sound and politics in an academic setting. We stopped at different points during the walk, following the score instructions to help frame our relationship to the educational space. The soundwalk was executed with a written score (Figure 6.1), which prompted audiences to reflect and listen to the political dimension of the university soundscape. The score was printed as a postcard and translated into Spanish, French, and English.

"LISTENING/SPEAKING IN TONGUES"

A SOUNDWALK BY AMANDA GUTIERREZ

SOUNDWALK SYMPOSIUM AT CONCORDIA UNIVERSITY

STOP 1

Take two pencils and tap them, matching the pulse of everyone in the group. Resonate with the space.

STOP 2

Think and whisper the names of BIPOC professors, colleagues, scholars, friends, and partners who influence your practice.

STOP 3

After you walk away, find a place on campus where you can listen to the surrounding voices, steps, and echoes of this building.

FIGURE 6.1 Listening/Speaking in Tongues (2022) score by Amanda Gutiérrez

During the soundwalk, I read excerpts of Anzaldúa's literature with a critical perspective on hegemonic Anglo academia. On the first stop, we started by reading the land acknowledgment to situate ourselves in the land of Tiohti:áke, now known as Montreal. Beginning with the recognition of our settler relations with space opened the opportunity to practice critical listening positionality, grounding the audience in the university environment. The listening exercise tuned our ears and guided participants into thinking about access to knowledge and education issues in Canada. The second stop was at the Henry F. Hall Building, where I read an excerpt of Anzaldúa's "How to Tame a Wild Tongue" (in Anzaldúa, 1987/1999), next to the busts of Concordia's previous presidents. The reading asserted a critical comment on taming a language as a strategy of cultural assimilation and gender codification.

The prompt of this stop guided participants to listen to the soundscape hall, highlighting the presence of different languages spoken by immigrant students. The second stop led to the indoor escalators, listening to the electric mechanism resonating as it went upstairs to the Computer Center on the ninth floor. This stop framed the historical event of "The Sir George Williams Affair," a student riot in 1968, denouncing discrimination and systemic racism at Sir George Williams University, now known as Concordia.[4] In the ninth-floor hallways, the soundwalk invited participants to imagine the sonic tumult of the student occupation, their meetings, the soundscape protest on fire, and the final arrest of students who participated in the action. Situated listening gave space for reflection and silence, allowing the audience to speak and share their historical knowledge. In this section of the walk, one student member of the BIPOC Student Association delved into the event's details and helped us to contextualize the riot's political relevance in the resonance of transnational Black protest in Montreal (Cummings and Mohabir, 2021).

The last stop recalled the life and death of Gloria Anzaldúa, which could have been prevented if her academic job had provided employment benefits such as health insurance. A fact that opened a critical conversation on the structural oppression and exploitation of unfair labor conditions that women of color experience in academia. As an ending, the walk invited participants to leave the group and walk alone to find a space to listen to the university soundscape.

Whispering Gloria Anzaldúa (2022)

The use of technology has a critical relation to my soundwalks, primarily pointing out the aural border thinking implication with space and virtual sound. *Whispering Gloria Anzaldúa* (2022) is a soundwalk designed for Echoes, an Augmented Reality (AR) platform based on a geolocative sound system. This soundwalks series considers an AR spatial sound composition that expands political listening and group discussions.

This AR audio walk features extracts from Anzaldúa's book *Borderland/la Frontera: The New Mestiza* (1987/1999) in connection with sound recordings of Mexican diasporic soundscapes of Chicago and Brooklyn. The soundwalk composition employed the aural border thinking method to experience comparative listening of Mexican immigrant soundscapes in these two cities. The composition amplification and storytelling respond to the site-specific acoustic and pedestrian context while emphasizing the symbolic relation with migration. I amplified the soundtracks using a wireless speaker connected to my phone and the Echoes app. Echoes tracked the sound path geolocated with augmented sound, activated through walking—the soundwalk design layers out Anzaldúa's text recordings, virtual soundscapes, and the site-specific aural environment. The silences allowed us to listen to the space while sharing reflections on aurally situated experiences on the concept of the borderlands. In this walk, critical listening positionality works with storytelling and virtual soundscapes juxtaposed in a physical environment. The main goal is to employ listening and dialogue as an aural exchange between positionalities, considering walking as a tool to provoke cultural dissonance among participants prompted by Anzaldúa's poetry:

> Cuando vives en la frontera
> people walk through you, the wind steals your voice,
> you're a burra, buey, scapegoat,
> forerunner of a new race,
> half and half-both woman and man, neither—
> a new gender;
> To live in the Borderlands means to
> put chile in the borscht,

eat whole wheat tortillas,
speak Tex-Mex with a Brooklyn accent;
be stopped by la migra at the border checkpoints.
(Anzaldúa, 1987/1999, p. 194)

Even if all the AR walks have the same audio content, each geolocation sound composition differs because it considers the site's geography, local context on migration, and the audience's response, drastically changing the soundwalking experience. The soundwalk was premiered for the first time in the neighborhood of Villeray in Montreal in 2021. The second iteration was at Governors Island in New York, and the third one in Pilsen, a Mexican American community in the city of Chicago; both events performed in 2022. These last events happened during the day as group walks in public parks with busy and casual affluence. Therefore, these events were planned as an open invitation to pedestrians in the park field, attracting people without previous experience of soundwalking, who followed us casually and quietly until the end. The event in Montreal happened at night and emotionally affected a few attendees whose identities were immigrant, Queer, Black, and Latinx. Their responses and expressions were emotionally charged, which gave a personal dimension to the group experience and verbally responded to the prompts from a subjective level. I believe that the intimacy of the setting, the closeness of our relations, and the symbolic bridge with Anzaldúa's poems prompted a space to share our vulnerability as a point of connection.

Walking in Lightness (2018)

The script of the film essay *Walking in Lightness* (2018) reflects, through auto-ethnography, experiences of accessibility by walking and listening to social cues in the public space. I recorded these sonic observations in two predominantly Mexican immigrant neighborhoods, Sunset Park in Brooklyn and Spanish Harlem in NYC. The video essay makes an analogy using the darkroom photo print process as the adaptation mechanism in a foreign culture. The video features two voices, one in Spanish and the second in English, reflecting on the inner experience of invisibility and visibility when asserting an identity as an immigrant. The video soundtrack is composed of sound recordings of Sunset Park and Harlem. The sound recordings are a subtle observation of embodied interactions during the walk. However, even after I recognized the sonic cues from Mexican culture, I could not read the aural complexities of the space just by walking and recording.

To position myself as a listener and an immigrant settler in this project, I taught English as a Second Language class as a teaching artist for the *NURTUREart Residency*, as part of their outreach program in collaboration with the elementary public school *PS136* in Sunset Park, Brooklyn, NY. The class curricula focused on aural pedagogies engaging with the cultural diversity in the neighborhood while inserting English language immersion. The 5th-grade class reflected the diversity of ethnicities and races in the school and their community. All students were recent immigrants, mainly from Global South countries, who were required to take the ESL (English as a second language) class to gain linguistic skills and assimilate to their guest culture. However, the class professor and I planned on first asserting and listening to the languages spoken in the classroom and the cultural sonorities in Sunset Park. Through several soundwalks, maps, drawings, and written exercises, we developed vignettes that imagined their ideal public space in their neighborhood and acoustic environment, considering what already resonates in the location. They recognized Spanish, Mandarin, and Persian, among other linguistic codes and accents, embedded in the rhythmical music from specific regions and the sonic landmarks in the street (Figures 6.2–6.3).

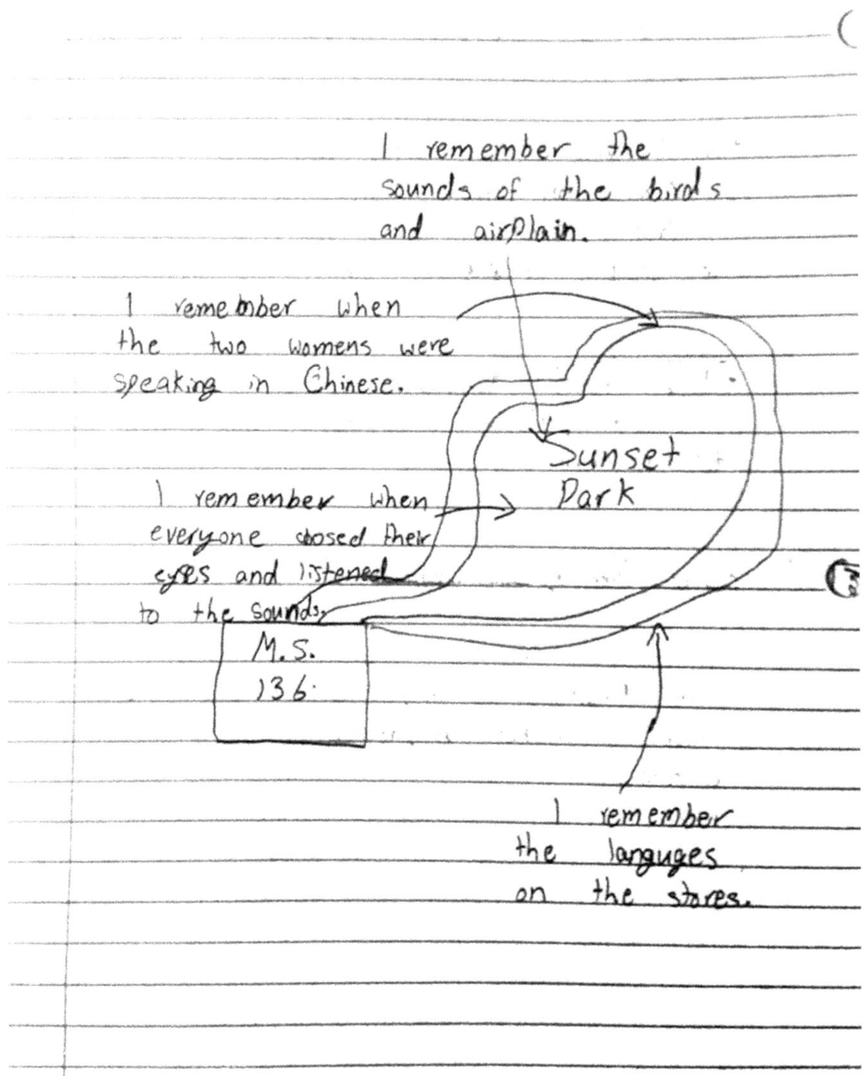

FIGURE 6.2 Sound map of a student auditory experience of Sunset Park

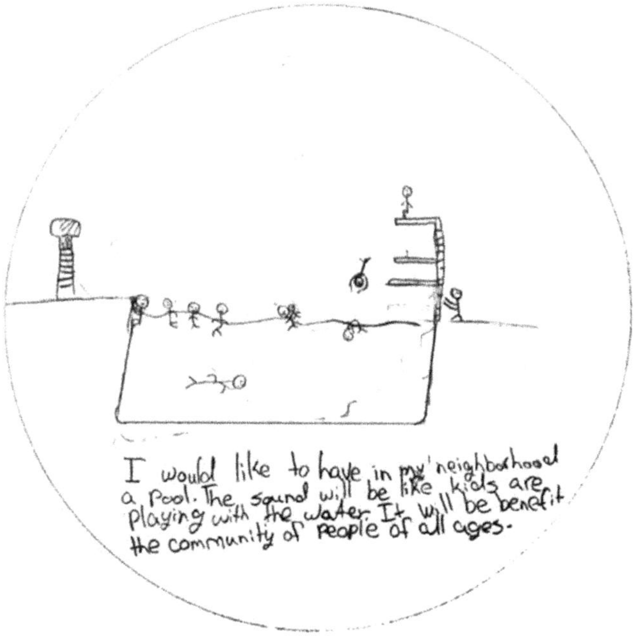

FIGURE 6.3 Student vignette of an imagined soundscape

We heard the screaming of the garbage man announcing the trash pickup, the street food vendors' bell attached to the "paletero" (ice pop vendor) cart while recognizing a language, an accent, or overhearing conversations. We heard and identified the music from different regions in Mexico and Puerto Rico, and the Tex-Mex music amplified with speakers in the streets to attract customers to their business. The aural observations guided our projects that highlighted the complexity and richness of the immigrant cultural soundscape. We concluded and argued that even if Sunset Park is culturally diverse, it is still segregated and needs a better public space infrastructure for younger residents and their families.

In Anzaldúa's literature, her observations are vivid aural memories that are part of her critical approach to multiculturalism. Her work can serve as an example to prompt situated listening exercises conveying cultural and embodied relationships with sound:

> The whole time I was growing up, there was norteño music sometimes called North Mexican border music, or Tex-Mex music, or Chicano music, or cantina (bar) music. I grew up listening to conjuntos, three- or four-piece bands made up of folk musicians playing guitar, bajo sexto, drums and button accordion, which Chicanos had borrowed from the German immigrants who had come to Central Texas and Mexico to farm and build breweries. In the Rio Grande Valley, Steve Jordan and Little Joe Hernandez were popular, and Flaco Jimenez was the accordian king. … I remember the

> hot, sultry evenings when corrido—songs of love and death on the Texas–Mexican borderlands—reverberated out of cheap amplifiers from the local cantinas and wafted in through my bedroom window.
>
> *(Anzaldúa 1987/1999, pp. 61–62)*

These embodied observations open a poetic dimension of a *fronterizo* identity (Anzaldúa, 1987/1999), which permeates her writing. Her words point at the complexity of the musical immigrant culture colliding in the borderland soundscape. I associate this excerpt as an example of the aural border thinking through storytelling achieved by situated listening.

The literature of Anzaldúa opens the possibility of situating ourselves in writing and storytelling through *autohistoria*[5] (Anzaldua 1987/1999), a concept with the interdisciplinary potential to cross creative writing and storytelling fields. This literary figure considers *La facultad mestiza* (Anzaldúa, 1987/1999) as the agency given by social consciousness existing among immigrant cultures and borders, where identity hybridity is an essential part of being mestiza.[6] As Anzaldúa recalls, this book opened a new dimension to understanding the term mestizaje:

> In *Borderlands*, I wanted to give that kind of accessibility to mestizaje, to open up the term, not just for Chicanas or women of mixed heritage, but to everybody who is limited by the nationalistic impulse. The lines and borders between races have already been transgressed and crossed, so there is no pure category that any of us can be put in and nailed.
>
> *(Hernández and Anzaldúa, 1995, p. 9)*

From a decolonial feminist perspective, it is essential to consider Maria Lugones' reading on consciousness implicit in Anzaldúa's *La facultad mestiza* concept, where she highlights the break of boundaries as an entry point to create collectivity as a political alliance:

> The mestiza consciousness is characterized by the development of a tolerance for contradiction and ambiguity, by the transgression of rigid conceptual boundaries, and by the creative breaking of the new unitary aspect of new and old paradigms. The mestiza consciousness participates in the creation of a new value system through an "uprooting of dualistic thinking" (Anzaldúa, 1987, p. 80). La mestiza is captive of more than one collectivity, and her dilemma is which collectivity to listen to. She crosses from one collectivity to the other and decides to stake herself in the border between the two, where she can take a critical stance and take stock of her plural personality.
>
> *(Lugones, 1992, p. 34)*

La facultad mestiza is translated in writing by asserting self-knowledge as a relational approach to self-writing (Pitts, 2016), which Anzaldúa names *autohistoria*. In the case of immigrant subjectivities, the *autohistoria* carries the opportunity of comprehending identities in constant flux between genders, borders, languages, classes, and cultures, which can be manifested through an intersectional and interdisciplinary creative process.

The aural border thinking methodology incorporates self-knowledge through *autohistoria* as an autoethnography tool. The autohistoria can be applied in sound practices through storytelling, actions of performative and situated listening. My argument is that autohistoria is a storytelling method that can be implemented in soundwalking practices through aural border thinking as a dynamic tool of self-reflection and social resonance. This mode of listening proposes a dialogue to understand the complexity of the migrant, settler, and indigenous roots of experiences that we carry as individuals. The soundwalk then incites listeners to hear specific cues in the site-specific soundscape to engage with conversations centered on the participant's personal experiences. The autoethnographic approximation of autohistoria in political writing and narrative-based artworks allows naming social oppressions through *speech and action* (LaBelle, 2018).

The following section will reflect on the political dimension of actions that employ sound and walking as tools of pronunciation in the public space, considering political movements and social relations in the borderlands.

The political dimension of collective sounding

LaBelle's concept of *speech and action* aligned with collective listening reinforces political resonances of sound subjectivities. The spaces where these discourses happen open a forum for personal exchanges to assert critical listening positionality in the public space:

> Questions of contemporary social conditions, and what may constitute public life today, as well as how the "space between" people may open up to enable formations of not only public discourse but equally gestures of joining together, become central to reconstituting the political realm. This is additionally shaped by an understanding of "speech and action" as being grounded in sounded subjectivities, where speech and hearing, voice and listening, form the essential and enduring means by which to nurture spaces between, especially when such spaces must contend with contemporary systems that impose a "privatization" to our senses and politics.
>
> *(LaBelle, 2018, p. 11)*

The "space in between" opens an opportunity to listen to self-referential positionality when creative modes of collaboration happen, especially in the public space. The streets manifested a space for subaltern subjectivities and aural practices informing political actions. These platforms and networks consider individual experiences while asserting mutual care through *performative assembly* (Butler, 2015) and political emancipations. Sonic emancipatory practices are essential in self-organized initiatives seeking and achieving autonomy, approaching sound as a political tool to assert a *space of appearance* (Arendt, 1958), which aligns political actions of self-representation in the public sphere. Appearing as an act of listening and sound creation brings possibilities for solidarity based on aural gestures. These sonic alliances can happen by exploring tools of collaboration that enable political agencies through aural pedagogies. These auditory tools are explored in the *militant sound investigation* (Ultra-Red Collective, 2008) as an ethical approach to exercising sonic activism. The manifesto *10 Preliminary Theses for Militant Sound* (Ultra-Red Collective, 2008) provides a framework to work as initiators in alliances, which is defined as inter-subjective participation (Ultra-Red Collective, 2008) manifested as an open agreement of collaboration in political executions.

Political sound executions happen through emancipatory practices, understanding cultures of protest as emergent coalitions. Walking in connection with sound can be manifested, forming communities in movement (LaBelle, 2018) as emerging social spaces without a set location but as itinerant collectivities. An example of these communities is the refugee political actions of the *Caravan of Mothers of Missing Migrants*, which had walked from South to North America, searching for disappeared family members and demanding substantial changes in the migration policies.[7]

> Carrying their disappeared family members' photographs around their necks, las Madres, who come from Honduras, El Salvador, Guatemala, Nicaragua and Mexico, demand justice and call out those responsible, including cartels, gangs and state forces. They march down bustling city streets and tranquil colonias, making their presence known by chanting, "Vivos se los llevaron, vivos los queremos" (Alive you took them, alive we want them) and "Hijo, escucha, tu madre está en la lucha (Son, listen, your mother is in the fight)."
>
> *(Sostaita, 2019, p. 3)*

These caravans execute emancipatory practices of social power. Their collective walk opens a space of appearance that challenges physical and symbolic borders through their vocal clamors. Acts of listening strengthen their organization through personal dialogues on their situated experiences, expressed in the marching lines among protesters and the street witnesses. These itinerant spaces enable social solidarity towards one common objective, finding disappeared family members and for migrants a safer dwelling to exist.

For the larger immigrant caravans from South America to the Mexican United States Border, migrating to a safer location is a survival goal that informs personal alliances and political coalitions. In forming those spaces, migrant settlements build their sonorities that carry linguistic codes, sonic cues, and sound landmarks that identity and inform their community formations. The sonic dimension residing in the borderlands is recorded and analyzed by sound artist and scholar Alvaro G. Díaz Rodríguez. His research paper "Sound through the Looking Glass. An Approach to the Dimensional Sonology on the Tijuana-San Diego border" (Díaz Rodríguez, 2020) delves into the aural dimension given the sociological sonorities on the border as well as in migrant and refugee camps in Tijuana. He proposes a *multidimensional listening* (ibid.), which considers the analysis of the geographic, economic, and social dimensions implicit in the soundscape. He concentrates his research on the aural study at the migrant camp, "El Chaparral." In his recordings, he conveys the relationship between Central and South American immigrants sharing the same space with Haitian immigrants. Díaz Rodríguez's sound work highlights the complexity of the geopolitical relationships inside the camp. The sonorities recorded point to complex systems of power, negotiation, and identity, audible in the camp's common spaces such as the kitchen, the bathrooms, and the central patio. Diaz's sound recordings capture intimate glimpses of the quotidian inside El Chaparral, employing ethnographic methodologies. The sounds assert multidimensional listening, which pays special attention to social interactions (Díaz Rodríguez, 2021).

These examples represent a small glimpse of the complexity of the topic of migration employing sound as a political tool of awareness and visualization. Both instances, by enacting emancipatory practices or while making public through sound studies, highlight the social and political experiences of migrants on the border of the United States and Mexico. The same borderland that Gloria Anzaldúa reflects as a first-generation immigrant considering herself as a feminist Chicana who found these connections through a border thinking process.

The concept of border thinking is approached by decolonial feminist scholars in the configuration of creating a feminist transnationality. The border in their conception is not a split or perpetuation of dichotomous ideologies but a space of resistance to the "coloniality of gender" and response to "colonial difference" (Lugones, 2010). For María Lugones, decolonial feminism is to learn from each other as resisters, which can be applied in the concept of aural border thinking by understanding the local thinking of sound as resistance—thinking as a coalition and weaving histories of resistance from all our situated knowledge. According to Lugones, community resistance to coloniality should be experienced since individualism drives isolation. The collective narratives of experience and the local and personal subjectivities as "beings in relation" are essential to cement a coalition. Lugones quotes Audre Lorde, who proposed in her feminist theory that "the ethics of coalition in the making in terms of both being in relationships that extend and interweaves its people ground" (Lorde, 2007) and makes a connection with Chela Sandoval's proposal on social erotics and creative relations. Maria Lugones evokes both authors in the quote:

> *The responses from the fragmented loci can be creatively in coalition*, a way of thinking of the possibility of coalition that takes up the logic of coalition of feminists of color: the social erotics (Sandoval 2000) that takes on the creative, that permits enactments that are thoroughly chotomies (Lorde 2007). The logic of coalition dichotomies; differences are never seen in dichotomous has as its opposition the logic of power.
>
> *(Lugones, 2010, p. 755)*

Lugones brings the practical concept of creating a ground for dialogue inside and outside academics, as well as reflections that are meaningful in applicable terms, which can frame an aural border thinking methodology:

> How do we learn about each other? How do we do it without harming each other but with the courage everyday that may reveal deep betrayals? How do we cross without taking over? With whom do we do this work? The theoretical here is immediately practical.
>
> *(Lugones, 2010, p. 755)*

These questions reflect authorship and collaboration, considering shared knowledge as the terrain and space for social and gender consciousness. The border thinking and the mestiza consciousness are essential concepts to consider while creating a collaborative *transfronterizo* space for the diasporic thought on experiences of coloniality.

Conclusions

My practice on decolonial soundwalking focuses on the auditory reception of personal narratives in connection with soundscapes. Their translation into embodied practices implies walking, applying aural pedagogies that involve sensorial orientations to the self. These practices aim to evolve into methodologies based on critical listening positionality and aural border thinking as tools to approach the social dimension of spatial sonorities in the context of migration. I guide these observations into interdisciplinary practices that listen to site-specific environments and storytelling. The positional listening that I propose feeds from the decolonial theory of Gloria Anzaldúa, María Lugones, and Dylan Robinson, which prompts participatory engagement in interdisciplinary sound and self-writing (autohistoria) practices. These methods aim to assert a multidimensional listening implicit in walking and invite participants to execute self-reflective positionality.

Participation in my work involves the possibility of engagement through workshops or arts and education practices, resulting in collaborations to develop the itinerary, route, and content of the walk. Otherwise, when I design and lead a soundwalk, I plan situations to break the silence and engage through speech and auditory actions. Then listening is focused on the soundscape and the situated experience of people participating in the soundwalk. Challenging and finding new perspectives apart from the canonical soundwalk procedure of listening in silence is part of the decolonial process. Processes of aural decoloniality urge other forms of engagement besides the Western aesthetic experience of solitary contemplation. Emphasizing that sounding with the environment has existed in non-Western cultures and engages with collective storytelling as their primary form of incarnate spatial experiences.

In the task of capturing field recordings when exercising a walk alone or collectively, it is essential to question the interpersonal dynamics established with the environment and its inhabitants. The recognition of our role and positionality is critical in documenting spatial sonorities. While framing reflections on migratory experiences, I keep in mind the complexity of these power dynamics that exist by representing subaltern subjectivities, especially in artistic or academic settings. The awareness of these dynamics is crucial in determining ethical relations with participants as project collaborators in the walk. The main goal in the inter-subjective participation points to assert a space of appearance where collective walking embraces political visibility. The main goal is to achieve itinerant and collaborative speech experiences and action of site-specific narratives in public space. The process of creating a soundwalk collectively should engage with radical pedagogies, considering methods of situated storytelling and multidimensional listening.

In these interactions, we must keep in mind whom we are speaking with and for. Who is invited to walk and who is to co-lead the walk, how do we determine the path, and plan the participatory prompts that will involve the soundwalk experience? Most importantly, do these conversations flow back to marginalized communities experiencing social oppression in the locations where we walk? The awareness of our role in socially engaged projects is vital in relationships with the communities we collaborate with in employing relational accountability as an ethical form of research creation (Wilson, 2008). Identifying our positionality will inform the use of aural pedagogies and methodologies needed to achieve critical thinking and mutual understanding. These are reflections through movement while delving into interpersonal memories with a site.

The soundwalk as a situated practice fosters listening acts that strengthen political solidarities by speaking, hearing, and reflecting upon our ancestors, grandmothers, and aural points of origin. To break and challenge hegemonic narratives of coloniality. By doing so, we are opening a space of listening where *otros mundos son posibles*/other worlds are possible while soundwalking collectively.

Notes

1 Self-awareness and reflexivity are key to name while writing about immigrant subjectivity since "self-referential positionality should also become a more explicit focus in intersectionality research, since various subject positions may or may not change dramatically over time (change of sex, sexual orientation, age, mental and physical health, level of education, social class, marital status, citizenship status, etc.), affecting what and how a presumed 'same' person remembers and forgets about their earlier subject positions" (Simandam, 2019, p. 137).

2 "Deslenguada's" literal meaning in Spanish is the female adjective of someone who is tongueless. The concept of "deslenguada" by Anzaldúa refers to the symbolic experience of missing a mother language. She refers to the Spanish as a Chicana in the United States.

3 A compilation of these works was published in a book titled *This Bridge Called My Back: Writings By Radical Women of Color* (1981/2015), edited by Gloria Anzaldúa and Cherrie Moraga. This book waves provocative essays, poetry, interviews, creative writing with visual arts and photo documentation of performance art. It opens a robust framework for interdisciplinary works voicing statements on racism, class, and gender oppression by emerging writers and artists of marginalized communities in the 1980s.

4 The book *The Fire That Time: Transnational Black Radicalism and the Sir George Williams Occupation* is a significant source edited by Ronald Cummings and Nalini Mohabir. The book provides historical research on the political legacy of this event in the current student movements in Montreal. For more information on its oral history, listen to this podcast: https://www.cbc.ca/radio/sunday/the-sunday-edition-for-february-3-2019-1.4997146/how-the-sir-george-williams-protest-changed-the-conversation-about-racism-in-canada-1.4998442 (accessed July 21, 2022).
5 The meaning of the term "autohistoria" in Spanish language is self-story. However, Anzaldúa plays with the two words auto and historia, placing them together as one.
6 *Mestizo* is an ethnic/racial term to classify individuals who have a racial mix of European and Indigenous American. This was first defined during the Spanish colonial empire. However, in contemporary critical race theory, Gloria Anzaldúa used the term *mestizaje* as empowering miscegenation, understanding race as a complex mixture of ancestries.
7 To listen more about the organization and political actions of *Caravan of Mothers of Missing Migrants*, follow this podcast: https://remezcla.com/features/culture/caravan-mothers-missing-migrants-trave-mexico-demand-justice/ (accessed: July 30, 2022).

References

Anzaldúa, G. (1987/1999) *Borderlands/La Frontera: The New Mestiza*, 2nd edn. San Francisco, CA: Aunt Lute Books.

Arendt, H. (1958) *The Human Condition*. Chicago, IL: University of Chicago Press.

Butler, J. (2015) *Notes Toward a Performative Theory of Assembly*. Cambridge, MA: Harvard University Press.

Cummings, R. and Mohabir, N. eds. (2021) *The Fire That Time: Transnational Black Radicalism and The Sir George Williams Occupation*. Montréal; Chicago; London: Black Rose Books.

Díaz Rodríguez, A. (2021) *Serie Campamento de Migrantes en TJ*. Sound Recordings https://soundcloud.com/diazalvaro/sets/serie-migrantes-en-el (accessed March 8, 2022).

Díaz Rodríguez, A. (2020) Sound through the Looking Glass. An Approach to the Dimensional Sonology on the Tijuana-San Diego border. In: A. Díaz Rodríguez, ed., *Musicians' Migratory Patterns: American–Mexican Border Lands*. London and New York: Routledge, pp. 1–29.

Haraway, D. (1988) Situated Knowledges: The Science Question in Feminism and the Privilege of Partial Perspective. *Feminist Studies* 14, 575. doi:10.2307/3178066

Hernández, E., and Anzaldúa, G. (1995) Re-Thinking Margins and Borders: An Interview with Gloria Anzaldúa. *Discourse*, JSTOR 18, 7–15.

Icaza, R. (2017) Border Thinking and Vulnerability as a Knowing Otherwise. *E-International Relations*, June 9, 2017. Available at: https://www.e-ir.info/2017/06/09/border-thinking-and-vulnerability-as-a-knowing-otherwise/

LaBelle, B. (2018) *Sonic Agency: Sound and Emergent Forms of Resistance*. London: Goldsmiths Press.

Lorde, A. (2007) *Sister Outsider: Essays and Speeches*. Berkeley, CA: Crossing Press.

Lugones, M. (1992) On Borderlands/La Frontera: An Interpretive Essay. *Hypatia* 7, 31–37.

Lugones, M. (2010) Toward a Decolonial Feminism. *Hypatia* 25, 742–759. doi:10.1111/j.1527-2001.2010.01137.x

McRae, C. and Nainby, K. (2020) (Re)sounding Pedagogy: A Themed Issue on Critical Communication Pedagogies of/for/in Sound. *Review of Communication* 20(4), 287–297. doi:10.1080/15358593.2020.1826567

Mignolo, W. (2012) *Local Histories/Global Designs: Coloniality, Subaltern Knowledges, and Border Thinking*. Princeton, NJ: Princeton University Press.

Moraga, C., and Anzaldúa, G. eds. (1981/2015) *This Bridge Called My Back: Writings By Radical Women of Color.* 4th edn. Albany, NY: State University of New York (SUNY) Press.

Pink, S., Hubbard, P., O'Neill, M., and Radley, A. (2010) Walking Across Disciplines: From Ethnography to Arts Practice. *Visual Studies* 25, 1–7. doi:10.1080/14725861003606670

Pitts, A.J. (2016) Gloria E. Anzaldúa's *autohistoria teoría* as an Epistemology of Self Knowledge/Ignorance. *Hypatia* 31, 352–369. doi:10.1111/hypa.12235

Robinson, D. (2020) *Hungry Listening: Resonant Theory for Indigenous Sound Studies.* Minneapolis, MN: University of Minnesota Press. doi:10.5749/j.ctvzpv6bb

Simandan, D. (2019) Revisiting Positionality and the Thesis of Situated Knowledge. *Dialogues in Human Geography* 9, 129–149. doi:10.1177/2043820619850013

Sostaita, B. (2019) *The "Caravan of Mothers of Missing Migrants" Is Traveling Through Mexico & Demanding Justice for Their Children.* Remezcla. URL https://remezcla.com/features/culture/caravan-mothers-missing-migrants-trave-mexico-demand-justice/ (accessed March 8, 2022).

Sotelo, L.C. (2010) Looking Backwards to Walk Forward: Walking, Collective Memory and the Site of the Intercultural in Site-Specific Performance. *Performance Research* 15, 59–69. doi:10.1080/13528165.2010.539880

Ultra-Red Collective (2008) *10 Preliminary Theses on Militant Sound Investigation.* New York: Ultra-Red Collective.

Velez, E. and Tuana, N. (2020) Editors' Introduction: Tango Dancing with María Lugones: Toward Decolonial Feminisms. *Critical Philosophy of Race* (Project MUSE) 8, 1–24.

Wekker, G. (2018) The Arena of Borders: Gloria Anzaldúa, Intersectionality and Interdisciplinarity. In: R. Buikema, L. Plate, and K. Thiele eds., *Doing Gender in Media, Art and Culture: A Comprehensive Guide to Gender Studies.* London: Routledge, pp. 54–69.

Wilson, S. (2008) *Research Is Ceremony: Indigenous Research Methods.* Halifax, Nova Scotia: Fernwood Publishing.

7

PATHS OF DEPENDENCE: WELCOMING THE UNWELCOME

Tim Shaw

Introduction

Technologically mediated soundwalks became increasingly common during the early part of the twenty-first century. Mobile technology, locative media and the possibility afforded by portable, Internet enabled smartphones have made small digital devices more accessible and available to artists, designers and researchers (Wilken and Goggin, 2014). With this increased popularity we find a particular type of soundwalk emerging, one in which pre-recorded sound files are played back using media players or triggered via geo-located boundary markers or time stamps (Özkul, 2015).

In this chapter I argue that these types of soundwalks, though often interesting and satisfying endeavours, do not make available to audiences the process or method of making the work. Technological and compositional process are often removed from the listening experience. In recent years we have witnessed an amplification of technologically mediated artworks which provide immersive or seamless experiences for audiences. These kinds of works often execute a kind of blackboxing of the aesthetic experience (Latour, 1999). This way of implementing technology, whereby audiences are presented with a finished, fixed product is common across many forms of contemporary sound and media art (Shaw and Bowers, 2015). In my work as an artist, I am interested in revealing the technological procedure to make these processes available to audiences. It is important for me that the artistic method is open and aesthetically available for audiences to engage with, if they wish to. In an earlier project, through a collaborative artistic residency, I explored a way of engaging audiences in a technological process in a public setting in collaboration with a museum and music festival (ibid.). The walks presented in this chapter aim to highlight this issue and offer an aesthetic and technical repositioning to technologically mediated, walking-based art.

DOI: 10.4324/9781003193135-8

Drawing on three of my own soundwalking projects, I will present a method for conducting technologically mediated soundwalking. Each of these pieces attempts, amongst other things, to reveal the technological and compositional process of the walk to participants during the performance event. The three walks I describe are as follows:

- *Ambulation*, an extended soundwalk which uses various listening technologies to create an improvised performance with the immediate sonic environment. Audiences, who are walking with me, wear wireless headphones and receive a live mix of processed audio signals manipulated through various pieces of hardware and software.
- *Net Walk*, an extended soundwalk which broadcasts altered soundscapes and processed video to an online audience using Internet streaming technologies. Developed during the Covid-19 lockdowns of 2020 and 2021, *Net Walk* has become a method for sharing an embodied soundwalking experience to remote audiences.
- *The Rose Walks*, a performance made in collaboration with sound artist John Bowers involving the streaming of audio and video from a walk, back to a physical venue where another performer works with this material as part of an improvised performance in front of a live, physical audience.

The description and analysis that follows will open a space whereby technologically mediated soundwalks provide direct access for the audience to engage with the process, technological infrastructure and the making of the work.

FIGURE 7.1 Ambulation soundwalk in Lausanne, Switzerland as part of LUFF 2020. Image: Francis Gendre, 2020

Context

I will continue to use the term 'technologically mediated soundwalk'; by this, I mean any walk which uses technology to mediate the experience of sound. I am aware of the shortcomings and ambiguities surrounding words like *mediate* and *technology*. In some cases I would argue that shoes or societal systems are indeed technologies and involve mediation (Franklin, 1989). I would like to state that I am open to what listening technology can include (sea shells, listening horns, radios, headphones, inductive coils etc.). Technologically mediated soundwalks can take many forms and do not have to include digital or contemporary technology. There are many intriguing ancient examples of portable sonic media such as bullroarers or buzz bones which I would include within this definition.

In previous writings about soundwalking practices, I have chosen to use the term 'extended sound walking' (Shaw, 2020). By this, I mean soundwalks that extend the possibilities of listening through various technologies which increase perceptual reach (for example amplification, underwater microphones and electromagnetic receivers). I will also use this term to describe the soundwalks presented in this chapter.

Contemporary examples of technologically mediated soundwalks include geo-located sound maps, such as the ones available and developed through the Echos[1] platform, allowing creators to define boundaries whereby fixed sound files can be triggered and processed by geo-locative sensors on a mobile device. The participant in this context becomes a temporary human cross-fader; their movement slowly changes the volume of a pre-recorded sound file. Online sound archives, such as the ever-growing Radio Aporee,[2] allow listeners to hear user generated field recordings from all over the globe. The site even offers a 'geomixer',[3] which navigates listeners through a line drawn across the map, virtually transporting them to various places around the world. With the increased availability of radio transmitters and possibilities for portable streaming, artists can now transmit sound from their local environment via various broadcast technologies (including wireless headphones and the possibilities for streaming online) (Shaw, 2020). These kinds of walks have been developed by artists and researchers and have been reported on widely in conferences and journals over the last 20 years (See Gaye et al., 2003; Carras, 2019; Allison and Dell, 2012).

For Hildegard Westerkamp, the soundwalk is an event which facilitates a listening experience, allowing us to 'become aware of our immediate acoustic environment' (Westerkamp, 1974). The argument I develop through this chapter is that many technologically mediated soundwalks do not facilitate the awareness of the immediate; in fact they often obtrude the proximate sonic environment by playing back sounds recorded or composed in another space and time. In the same vein, what is appealing to me about soundwalking is that it creates a heightened awareness of environmental sound, allowing the outside in (Westerkamp, 1974). The complex soundscape thrusts itself upon the event of a soundwalk; many elements of the soundscape are indeterminate, created by other beings and actors who share the physical and sonic space. Soundwalkers can control the speed, duration and route of the walk but many other aspects lie outside of the listener's direct control. Unlike the traditions of electro-acoustic music or soundscape composition, soundwalking does not have to be made in a studio and presented in a performance space or via a recorded release. The compositional process and the performance of it happen simultaneously. Traditional roles of performers and audience are flattened, or put into question. To quote Westerkamp again: 'Listening implies a preparedness to meet the unpredictable and unplanned, to welcome the unwelcome' (Westerkamp, 2015).

To be able to welcome the unwelcome, we have to be open to it and create systems, technologies and events that facilitate opportunities to allow the outside in. Welcoming the unwelcome is not possible if we do not allow the unwelcome into our technological or listening systems. The walks described in this chapter attempt to extend Westerkamp's idea by welcoming the unwelcome. Aspects which are often avoided or ignored in technologically mediated soundwalks are brought to the fore. These include failures of technical infrastructure, materiality of networked technology, mistakes from extended field recording techniques and the characteristics of the media device and headphones. In these projects I aim to have an open and improvisational approach, attempting to flatten the process and presentation of the artistic activity within the performance event. I believe that the technological process should be (a)live and available to audiences. When we creatively engage with failing technological infrastructures, for example, we begin to understand these systems in different ways. The system ceases to be an invisible, seamless entity and begins to be something the artist and audiences can work with productively.

This deep engagement with technology allows for its material character to come through. The technology here becomes an integral part of making the work, rather than just an end product of the artistic process. By allowing technological systems to interact and interfere with the environment surrounding the work, we are given new perspectives on the environment, the technology and the practice of soundwalking.

Oliveros' 'Deep Listening' approach moves away from Western classical music traditions to consider environmental sound and holistic notions of listening as a creative act (Oliveros, 1996). Her written scores, such as *From Unknown Silences* (Oliveros, 1996), engage with the idea of sound and silence as structural forms rather than addressing rhythm, harmony or melody. The primary activity for the participant in Oliveros' work is listening, and through listening the decoupling of musical practice from virtuosic technique or harmonic theory to emphasise an engagement with sonic phenomena instead. The Deep Listening Institute promotes a 'heightened awareness of the sonic environment, both external and internal, and promotes experimentation, improvisation, collaboration and playfulness'.[4]

Within her deep listening practice Oliveros also engaged with soundwalking, and a number of her text scores include instructions for walking. For example, from her *Sonic Meditations* series:

> Take a walk at night. Walk so silently that the bottoms of your feet become ears.
>
> *(Oliveros, 1971)*

Oliveros' deep listening scores and her theory of sonic awareness resonate with my own experience of field recording and soundwalking. Her approach to composition integrates the phenomena surrounding the performance event. In her piece *Bon Feier*, for example, she invites performers to take part in mini-rituals and events in a city, college or university environment, following a word score.

> Special rituals, activities, and sights, … are to be blended smoothly with normal city or campus activity all during the normal working day and evening. The intention of *Bonn Feier* is to gradually and subtly, subvert perception so that normal activity seems as strange or displaced as any of the special activities.
>
> *(Beal, 2006)*

The structure of my soundwalks supports a collective act of listening through expanded recording technology. Similar to *Bonn Feier*, it directly engages with the phenomena of the spaces it is presented within. It is simultaneously a mediation of the performance event, and the performance event is a mediation of the place. Reciprocal relations between space and sound are written about by Nina Sun Eidsheim, who discusses sound always existing in transmission and never outside of it (Eidsheim, 2015). I appreciate Oliveros' idea of having an openness to the sonic elements outside of the composer's direct control. I believe listening to the technological infrastructure of mediated soundwalks is also related to ideas surrounding Deep Listening. We listen to the whole spectrum of sound, the hiss of the tape, the noise floor of the mp3, the wind as it passes through the headphones. Through a Deep Listening practice, we do not just listen through the technology, we are also invited to listen to it.

Janet Cardiff and George Bures Miller have worked together to create site-specific audio walks for numerous galleries, festivals and museums around the world. Many of Cardiff and Miller's walks require the listener to don headphones, carry a media device and follow directions given to them from a pre-composed recording. Often using binaural sound, the works include a voice over, field recordings and added sound design, and lead listeners through a constructed aural world (Cardiff, 2005). Cardiff describes her walks as a way of 'slowing down the process of telling a story' (ibid.), and many of her pieces rely heavily on a text-based narrative to drive the experience.

Describing the walks of Cardiff and Miller, Charles Stankievech argues that these soundwalks provide a complex interface between outside and inside listening. He names this phenomenon, *L'extimité*, an amalgamation of exteriority and intimacy.

> This strange orientation is not due to a loss of reality but to a supplement to reality. Nor does one become lost in reality. Instead listeners enter a hybrid state where they play a dramatic role that contrasts with everyday life and which is directly interpellated into their minds by the instructional voice of Cardiff.
>
> *(Stankievech, 2007 n.p.)*

This form of technological mediated soundwalk is very different from the soundwalks I present in this chapter. The composition and technological system are not presented as aesthetic components within the work. Here the media is fixed prior to the point of presentation. Though, as Stankievech comments, a curious hybrid reality is constructed through this listening space, the audience are removed from the making and technical process. These works are blackboxed and productised prior to the point of presentation.

Christina Kubisch is known for her 'electrical walks', which take a very different approach from that of Cardiff and Miller. In these walks Kubisch embeds electromagnetic audio inductors into headphones, which allow listeners to navigate their own way through an environment. Kubisch's extended soundwalks tend to take place in urban environments and encourage listeners to explore the electromagnetic fall out of cash machines, alarm systems, fluorescent lights and other electronic devices we find in our cities. This is a good example, by my definition, of both a technological mediated soundwalk and an extended soundwalk.

Kubisch's walks are a successful instance of a piece which welcomes the unwelcome. These electromagnetic waves, which are often byproducts of contemporary technical devices, are welcomed into the composition of the work. The specially made headphones provide a kind of filter, allowing walkers to listen in to sound worlds not usually available to their ears. My walks also involve a kind of filtering. My palette of technological listening devices provides me with a useful structuring element, allowing me a kind of compositional certainty when walking through diverse signals.

With the headphone walks conducted by myself, the immediately encountered sound is a live source of creative material; not framed within the recalling of memory but as an instant response to the environment. As such, rather than pre-constructed narratives these works are improvisations conducted with the uncertainty of the sonic world as it is encountered through the act of walking.

FIGURE 7.2 Ambulation soundwalk in the Saltern of Cervia, Italy as part of Elementi Festival 2022. Image: Chiara Pavolucci, 2022

Three walks

I will now provide detailed descriptions of three technologically mediated soundwalks developed over the last eight years. All of the walks are continuously in process and development. The nature of these projects is that they are never finished and are in perpetual evolution. The systems, technologies and tools are revisited with every invitation or incentive to walk. What I describe here are the key features and characteristics of each walk, not a definitive definition.

Ambulation

Ambulation is a soundwalk which has been in continuous development since 2014. With *Ambulation* I take the audience on a soundwalk whereby signals from the immediate environment are collected through various microphones and listening technologies and broadcast to a group of people walking with me. Using a small computer, I am able to live-mix, process and manipulate the signals we move through. I use a Pure Data[5] patch running on a BELA board computer.[6] The BELA board is controlled using a small midi controller which is attached to the top of my recording bag. The sound is broadcast to the wireless headphones worn by the audience via a compatible transmitter. The environmental sound is mediated through microphones, the recorder, the computer, the broadcast transmitter and the headphones. When introducing the walk to the audience, I explain the process and the system in detail. It is important to me that the audience is aware of the technological details of the work.

No pre-recorded sounds are used in the *Ambulation* soundwalk; all sounds heard by the audience are from the immediate activity of the sonic environment. The audience hears these sounds at the same time as I do.

When arriving in a new environment, I get to know the space through walking. Without any listening technology I wander through the area finding unusual and dynamic acoustic spaces. These can include shopping centres, busy streets, markets, train stations, pedestrian walkways, parks, forests, lakes, ponds, fountains, rivers, car parks and churches. A contrast of environments works best for an interesting walk. When I have decided on a route, usually lasting around 45 minutes, I walk it a couple of times to get familiar with it. Occasionally I deviate from the route, especially if something spontaneous grabs my attention during the performance. I do not use the *Ambulation* system until I perform it with the public. This means that the first time I am hearing the environments mediated through the system is when I am performing with the audience. At the end of a walk, I usually facilitate an informal conversation with audience members who want to stick around.

Ambulation embraces the uncertainty of the everyday environment. It welcomes the unwelcome. For example, when presenting *Ambulation* in Brisbane (Australia), locally sourced RF (radio frequency) headphones were so full of audio artifacts they became almost unusable. In contrast, in the city's Botanical Gardens, the interference of the urban environment did not impede the *Ambulation* signal as much. The interference experienced on walks such as Brisbane was an interesting problem and became something to tie in to the creative decisions when planning the walk. As *Ambulation* is about the experience of sound, including phenomena not usually within our perceptual reach, this interference felt relevant to explore rather than obstructive. Different species of 'interference' became incorporated into walks and became a part of the creative sonic material of *Ambulation*. The route in Brisbane started within the park where little or no interference occurred. Towards the end of the piece, the audience were taken into areas of interference where, in addition to the usual broadcast sonic material of *Ambulation*, various RF artifacts, audio cut outs and even the occasional taxi driver conversation could be heard. Though the exact character of such interference was beyond my control, a route can be shaped around how much or how little interference is desired and when. The walking route in these instances was built around the quality of possible broadcast, and moved through areas of 'compositional ambiguity', in which it would become unclear what was being performed and what was incidental to the walk.

At the time of writing, *Ambulation* has been performed at 34 festivals, events and conferences around the world. Being able to present this work in different environments has been very important to its development. The *Ambulation* system is one I can take to different contexts and perform with different audiences. The system and technologies change depending on the context I am invited to present the work within. This versatility of the work is important to me. It does not have a fixed, predetermined audience. For example, I have conducted *Ambulation* soundwalks to elderly people in a community centre, to eight-year-olds in a primary school, and after a nightclub programme at a festival. Though some sites are more successful for this kind of soundwalk then others, I am always interested in challenging myself to present this piece in different circumstances. Locations have ranged from ancient forests to brand new shopping centres, from heritage places to construction sites.

A full technical and aesthetic description of the *Ambulation* project can be found in an earlier paper (Shaw and Bowers, 2020). The system for the *Ambulation* soundwalk is available to download and modify for free via my personal website.[7]

Net Walk

Net Walk is an augmented soundwalk which broadcasts altered soundscapes and processed video to an online audience using Internet streaming technologies. Developed during the lockdowns of 2020 it has become a method for sharing an embodied soundwalking experience to remote audiences. In *Net Walk*, I take a walk with microphones and a digital video camera (usually a smartphone) and broadcast the streams of audio and video to the Internet. In this walk, sound and image are entirely mediated through these audio-visual technologies.

I take a walk alone, carrying a collection of listening and sensing technologies that allow me to collect and extract different elements from the immediate environment. I mix and process this material through a BELA board computer (the same microcomputer I use for *Ambulation*) and send the result to a smartphone-compatible soundcard connected to a smartphone. The phone is connected to a portable hotspot which allows the device to connect to the Internet via a cellular connection. Various Video Conferencing tools were experimented with during these walks. I was re-appropriating Zoom[8] for soundwalking purposes, which was a contrast to how it was used for many people in Europe during the lockdowns of 2020. On the other hand, expecting people to spend more of their attention on Zoom beyond the working day was a big ask. Twitch[9] ended up being a good alternative to Zoom as it was simple to connect to and could be easily embedded onto a webpage.

During the walks I build up and down the layers of the composition using signals encountered during the journey. Ambient soundscapes, electromagnetic energy and hidden resonances are collected and incorporated into the developing composition. Moving through different architectures and landscapes, the mobile Internet becomes patchy and sometimes completely unavailable. This meant that freezes, glitches and dropouts occasionally transpire. Though at first I considered these as failures, in time I decided to embrace these nuances as a way of creatively understanding the topology of the cellular networks of the various places I was moving through. The work embraced the technological infrastructure as an active agent. With this in mind, I started planning the walking route as a way of playing with this. Navigating through underground car parks or lifts would significantly reduce the strength of the cellular connection. In previous works I have explicitly worked with network latency and the limitations of broadcast media for creative purposes (for example Shaw et al., 2015). This project became an extension of those artistic interests within the context of a soundwalk.

As with *Ambulation*, *Net Walk* works with field recording as a live performative act. With *Net Walk* the audiences are not walking with me but rather access the work remotely from their own environments. It works with the limitations of the technology it uses, folding latency, dropouts and freezes into the piece. As the pandemic intensified and many cultural spaces remained closed during 2020, artists and musicians took to live streaming performances. The purpose of *Net Walk* was not to re-present a physical soundwalk to an online audience, it was to create a soundwalk especially for an online experience. *Net Walk* was developed as part of a residency at the Digital Art Studios in Belfast, UK. *Net Walk* has been presented eight times in collaboration with We're All Bats and The Walking Festival of Sound both in 2021. Documentation material can be found via the dedicated project website.[10]

The Rose Walks

The Rose Walks is a collaborative project between artist John Bowers and myself. It was presented in the Østra performance venue in Bergen as part of Piksel festival in autumn 2021. During the performance I took a walk around the streets of Bergen, starting from Østra, and streamed processed video and sound back to the performance venue. The video was projected onto a large screen at the back of the space. The sound was routed through a mixer and played through the venue's large sound system. Bowers was situated in the performance space and could process the incoming sound using a collection of feedback networks he had hand coded. Sound from the outside world was mediated through my walking system and broadcast to the performance space. This sound was simultaneously re-mediated through Bowers' feedback system, responding to changes in the broadcasts soundscapes and the acoustics of the performance space.

We began the walk with an introduction from inside the venue; my performance system was already broadcasting video. I explained to the audience the details of the system and what they could expect to see and hear as I walked around the city streets that evening. John then briefly introduced his performance system and talked the audience through the binaural listening head that the room sound would be processed through. After the introduction I lifted the fader of the streamed audio, and because of the latency you heard the final sentence of explanation being amplified through the main sound system. As this was then feeding into the microphones a feedback loop emerged.

I then left the performance venue and took my journey through the busy streets of Saturday night in Bergen. I pushed through crowds of people stood outside bars, past Kebab shops and city traffic. I picked up ambient sounds, electromagnetic fall out and the resonance from street objects. There were a few dropouts during the performance, once due to signal loss because of the connection of the hotspot and once due to a drunk person unintentionally pulling out the power cord for the smartphone broadcasting the video. During this time the screen in Østra either froze or went blank. After the show people commented on the dramatic effect this created; when the screen froze, people speculated about where I might be when it unfroze. As I walked the streets of Bergen, John provided ambient sonic support, creating textures and gestures using some of the places I walked through.

The performance concluded with me returning to Østra. As the audience could see where I was with the projected camera feed, there was a small group of people looking eagerly towards the door as I entered the upstairs room. Gradually, due to the latency, the feedback loop returned. This accumulation of noise gradually intensified and eventually became the end of the performance. *The Rose Walks* is an exploration of a soundwalk in connection with another, non-walking performer. Dropouts, freezes and latency caused by the technological infrastructure are incorporated into the performance aesthetic and become part of the creative material.

FIGURE 7.3 Ambulation soundwalk at Lokal-int art space in Biel, Switzerland. Image: Chri Frautschi, 2021

Observations and reflections

What unites these three walks is that they all allow the technological system to incorporate uncertainty and indeterminacy; no pre-recorded sound files are used in any of the walks. In each of the pieces, I hear the mediated sound at the same time as the audience; there is no pre-fade listening. The pieces unfold in the moment.

The soundwalking systems I have developed allow for a generative approach whereby sound, space and non-audible signals are folded into one another. Each walk is open and improvisational, and decisions are made on the spot. I do not have the composer's privilege of producing something in the safety and warmth of a studio. I am listening, recording and broadcasting the walk simultaneously with the audience. Technological interferences, handling noise, noisy peaks and mistakes are incorporated into the work. Sometimes the technological failures can be the most interesting element of the performance, and create new meanings and contexts for the pieces (for example the audio artifacts in Brisbane, as described earlier).

Below I describe four key findings which have emerged from the practice-research conducted through developing these walks. It is important to note that these findings emerged from the practice. I have discovered these themes through performing the walks multiple times in various contexts. These concerns were not preloaded prior to making the work.

1. The liveness of media

With the soundwalks presented here, I am interested in the media itself having a liveness. The liveness is made tangible in different ways. I am literally live-mixing between different live-signals; I am not conducting this activity in a studio prior to the event, but on foot, with the audience, within the environments we are listening to. There is also a liveness to the technological infrastructure. As we move through electromagnetic signals, I not only sense them using inductive coils but they also impose themselves onto the radio frequency headphones we wear. These signals also inform the 'quality' of the cellular connection in *Net Walk* and *The Rose Walks*.

These walks deal with the act of field recording as a performance activity. They engage with soundscapes as a live phenomenon, ever changing and geographically specific. The making of the work is conducted in close correspondence to where the sound originates, and does not involve the moving of material from one environment to another, but rather the processing of and engagement with sonic material in situ. Like Burtner's *Ecosonics* (2011), *Ambulation* engages with sound and site as part of the making process. The composition is not pre-made, but unfolds with the audience as a live improvisation. It also responds to the unpredictable nuances of the immediate environment.

This way of conducting technologically mediated soundwalks is very different from that of Cardiff and Miller, for example. Though there is a liveness to their work, with the pre-recorded file being unfixed by the unpredictable environment it takes place within, it does not incorporate the technological system as part of the work itself. Rather, in their work, the media becomes an invisible facilitator for the experience of recorded sound.

From the interferences in *Ambulation*, through the network dropouts in *Net Walk*, to the more human, cable-pulling drunk man in *The Rose Walks*, the chaos and unpredictability of the public space become accepted and incorporated into the structure of these works. The media is not a vessel which carries information from one place to another, rather it becomes a medium for understanding one's environment differently. The media here becomes part of the environment itself.

2. Situated improvisation and interfaces

The unpredictability of the sonic environment and its potential for musical expression are discussed through the soundwalks presented in this chapter. The walks are approached through a practice of improvisation; the uncertainty of the sound worlds are fundamentally tied into the pieces. Unlike the walks of Cardiff and Miller, which draw on fixed audio recordings and linear narrative structures, these walks work with the spontaneous soundscape as a creative, live agent. Improvising approaches to the media allows for unpredictable outcomes to occur. In his monograph *Improvising Machines*, Bowers discusses electroacoustic improvisation through 'responsive action', focusing on activities that fold in the productive characteristics of place, structure and technology. For Bowers these are productive features and not 'problematic obstructions' (Bowers, 2002). My soundwalks accept these unpredictable features and use improvisation as a *thinking through making* process (Ingold, 2013), extending Bowers' account of electroacoustic improvisation to the form of a performative soundwalk.

There is also a sense in which the technologies I am using are part of an *Aesthetic Interfacing* (Shaw and Bowers, 2020). The making of the work, the presentation of it, the thinking behind it and the perceptual-aesthetic experience are all intertwined. This offers a different orientation to 'site' from applications associated with locative media whereby pre-made sound files are abstractly mapped onto physical space. In the walks presented in this chapter the perceptual-aesthetic experience of the pieces emerges from multiple resources in juxtaposition, the site, the sounds, the technologies which are implemented, and the audience's attention. The interfaces here do not bring about a particular effect but rather work together to creatively respond to affective agents in the contributing soundscape. Working in this way demonstrates how sonic materials are composed with, translated and manipulated during the act of recording and how this practice blurs the line between studio and situated listening.

In his talk for The Walking Festival of Sound, Barry Truax describes soundwalking as a practice which 'combines the roles of composer, performer and the audience into one integrated activity'.[11] This flattening of roles offers new, interpersonal dynamics between those involved in the artistic activity of the soundwalks. Some technologically mediated soundwalks do not embrace this integration of roles. In these walks, the composition is made in a separate time and space and the artists are often not present at the point of presentation. In this type of walk the traditional role of the composer is maintained.

3. Designing for uncertainty

Artists have long worked productively with technological uncertainty. The experimental media work of Cage, Schwartz and Moholy-Nagy, for example, draws on randomness, change and unpredictability to create sound, video and print. Many technologically mediated soundwalks, do not incorporate uncertainty. When we facilitate the experience of technological failures within infrastructures, we begin to understand them differently. For anthropologist Brian Larkin, 'infrastructures produce the ambient conditions of everyday life: our sense of temperature, speed, florescence' (Larkin, 2013). When these infrastructures begin to break, they straddle a space between ambient experiences and more assertive ones.

In *Net Walk*, the cellular telecommunication infrastructure supports the streaming and reception of audio and video. Not only is the walk about responding to the immediate environment, but it also productively explores the interconnectedness of the infrastructures that surround that environment. When walking through the old town, with its narrow streets and densely packed walkways, we experience dropouts and freezes. Entering a lift in a train station provides a temporary Faraday cage, resulting in the electromagnetic waves of the cellular network not being able to permeate the hungry technology. Dark spots are not avoided but creatively folded into the composition of the walk. Engaging with these communication technologies, through a walking, streamed performance, gives us a deeper understanding of them.

As with most soundwalks, these pieces happen outside, and are subjected to the weather conditions of the environment we are walking within. I have performed these walks in all weathers and do not tend to cancel events because of rain or other weather. During one *Net Walk* performance, while it was raining, drops of water got into my MIDI controller and began to make chaotic connections between the electronic components. The fader sensor on one of the channels began to behave erratically. This resulted in a sound being faded in and out of the mix very quickly. Though at first this was a shock, I began to find it very interesting. The humidity from the rain was imposing itself onto the system of the walk. I began to improvise with and respond to this new autonomous aspect of the system. The uncertainty of the water within the electronics became its own element of the performance. It certainly played the fader in a way I was unable to. Electronic engineers will tell you that water is a circuit's worst enemy; it seemed here water had found a voice in my performance system. By letting the water in I was literally welcoming the unwelcome into my technological system. Thankfully, when I returned home and dried out the MIDI controller, the water ceased to speak.

These walks productively feature the uncertainty of the environments they are presented within. This ranges from the electromagnetic artifacts of the broadcast equipment to the metrological conditions of the performance environment.

4. Embracing anti-solutionist and unfinishedness approaches in technologically mediated soundwalking

In *Discourse on Thinking*, Heidegger talks of the different modes of human thought and specifies two types of thinking, meditative thinking and calculative thinking (Heidegger, 1966). For Heidegger, calculative thinking is concerned with planning, research and can be used to 'count on definite results' (ibid., p. 46). On the other hand, he describes meditative thinking as something which transcends problem-solution scenarios. It is 'in flight-from-thinking'; it takes its own path and functions within the limits of the individual's mind. Heidegger says that meditative thinking is no good for dealing with business or economic models. It cannot be used to carry out practical affairs, but it is essential to 'dwelling' in the world, to reflect, to take time and to ponder. Heidegger applies these models of thinking to our approach to technology. Meditative thinking 'enables us to keep open the meaning hidden in technology, *openness to the mystery*' (ibid., p. 55; emphasis in orig.). New meanings emerge from reflecting and meditating on our complex relationship with technical devices but also through non-solutionist approaches to technology. For Heidegger, meditative thinking must be practised alongside calculative thinking to allow our relationship with technology to 'flourish'.

My approach to the technological aspects of making these walks has affinities to Heidegger's meditative thinking. This has also been channelled through Ingold's writing in his distinctions between hylomorphism and morphogenetic ways of thinking (Ingold, 2013, p. 22). In *Ambulation*, alterations to the system, the choice of microphones and listening devices continued to change in different instances of the work. I respond to the material qualities of the space I was presenting the work within and reflected on previous instances of the walk in order to develop it. In *Net Walk*, limitations in the technology are foregrounded as creative materials of the work. The artwork attempts to reveal the hidden technological processes and to support a meditation on how it shapes the dynamic of the work. In all of the ways I have used technology, I move away from the idea of technologies being understood as mere tools. For me, technology is an artistic material, a material that can be approached through meditative rather than calculative thinking.

Many approaches to technologically mediated soundwalking take a calculative approach, whereby an idea is formed in the artist's head and technology is employed to realise it in the world. In *Pandora's Hope: Essays on the Reality of Science Studies*, Bruno Latour also discusses 'blackboxing' as a way in which 'scientific and technical work is made invisible by its own success. When a machine runs efficiently, when a matter of fact is settled, one need focus only on its inputs and outputs and not on its internal complexity' (1999, p. 304). Latour argues that it is through the opening of blackboxes of science and technology that we can understand the link between different elements of the technological process and the social dynamics that surround these technologies.

The walks described in this chapter work with the act of field recording as a live performance activity. Instead of simply moving sound material from one environment to another, these projects use recording technologies and signal processing to create a direct response to the immediate soundscape. Through a variety of self-built technologies and artistic strategies, I have developed methods for conducting technologically mediated soundwalks and approaching the collection of sound material as a live activity. The performance systems and sound technologies I use are constantly changing and are revisited in response to the environments I choose to present the walk within. This work uses field recording technologies to improvise within a shared soundscape. *Ambulation, Net Walk* and *The Rose Walks* approach the spatial and contextual character of sound in situ, and the liveness of this work comes from an engagement with the immediate environment in which it is presented. This places these works on a different terrain from that of other soundwalks, for example, whose narrative-driven sound works attempt to abstract and fictionalise the immediate environment through Foley audio and text-based 'storytelling'.

Conclusion

As described in this chapter, technologically mediated soundwalks are often presented in a way whereby the audience is separated from the process of making the walk or recording. I argue that these methods frequently separate making and presentation process, and the technology used to employ these walks is instrumentalised for particular effect, 'immersing' the audience in a particular virtual environment or creating specific boundary markers for the playback of geo-locative sound files. Authors of such walks commonly make invisible the artistic or technical process, creating clear distinctions between audience and performer, composer and listener. In my walks there is a *liveness to the media*, the media is not fixed or determined before the point of presentation.

In *Ambulation, Net Walk* and *The Rose Walks*, I respond directly to unexpected sonic material created as a consequence of the technological byproducts of urban activity. These artifacts are not limitations but have productive potential for the performance of the work. By adopting an open approach, through *situated improvisation*, this work adapts to environmental phenomena and change and remains responsive to the different places I present it.

The walks described address the central role that field recording has within my sound art practice by configuring it as a live, collective and perambulatory performance. Though the pieces work heavily with uncertainty and chance, they rely on a system which allows this uncertainty in. I have discovered, through multiple performances of these soundwalks, that *designing for uncertainty* is something key for presenting work of this nature. The technologies implemented into the work are in continual development. With these walks I am not presenting a finished, technical product to an audience, rather an unfinished and open system in process.

Though I do have control over the system, and I have composed the technological apparatus which make up the walks, I allow the components of the work to interact with elements of the environment outside of my control. There is a useful *unfinishedness* to the work which allows these elements in. These walks embrace technological failure as part of the procedural, aesthetic and perceptual aspects of the projects; they are an attempt to welcome the unwelcome.

This unwelcomeness also extends to the technological infrastructure surrounding the performance event, the radio frequencies of different countries, the environmental conditions of the performance space and the network topology of the built environment. The environment and its relationship with technology are intertwined, the surroundings enact themselves onto the technological infrastructure and vice versa. The soundwalk becomes an event which folds together listening, technology and environment, which walk together on a Path of Dependence.

Notes

1 https://echoes.xyz/ (accessed 20 March 2022).
2 https://aporee.org/maps/ (accessed 12 May 2022).
3 https://aporee.org/maps/work/geomixer.html (accessed 13 May, 2022).
4 https://www.deeplistening.org/ (accessed 11 February 2022).
5 https://puredata.info/ (accessed 18 May 2022).
6 https://bela.io/ (accessed 18 May 2022).
7 http://tim-shaw.info/projects/ambulation/ (accessed 11 July 2022).
8 https://zoom.us/ accessed (20 March 2022).
9 https://twitch.tv/ (accessed 20 March 2022).
10 https://net-walk.net/ (accessed 24 April 2022).
11 https://www.youtube.com/watch?v=h_Ih02cJ7lY (accessed 24 April 2022).

References

Allison, J. and Dell, C. (2012) AuRal: A Mobile Interactive System for Geo-Locative Audio Synthesis. NIME'12, May 21–23, 2012, University of Michigan, Ann Arbor.

Beal, A. (2006) *New Music, New Allies: American Experimental Music in West Germany from the Zero Hour to Reunification.* Berkeley, CA: University of California Press. p. 210.

Burtner, M. (2011) EcoSono: Adventures in Interactive Ecoacoustics in the World. *Organised Sound* 16(3), 234–244.

Bowers, J. (2002) Improvising Machines, Ethnographically Informed Design for Improvised Electro-Acoustic Music. Available at: http://cid.nada.kth.se/pdf/CID-195.pdf

Cardiff, J. (2005) *Janet Cardiff: The Walk Book.* Thyssen-Bornemisza Art Contemporary. New York: Walther Koenig.

Carras, C. (2019) *Soundwalks: An Experimental Path to a New Sonic Art. Organised Sound.* Cambridge: Cambridge University Press.

Eidsheim, N.S. (2015) *Sensing Sound Singing and Listening as Vibrational Practice.* Durham, NC: Duke University Press.

Franklin, U. (1989) *The Real World of Technology, CBC Massey Lectures.* Canada: CBC.

Gaye, L., Mazé, R., and Holmquist, L.E. (2003) Sonic City: The Urban Environment as a Musical Interface. In *Proceedings NIME*, Montréal, 22–24 May 2003.

Heidegger, M. (1966) *Discourse on Thinking.* New York: Harper & Row.

Ingold, T. (2013) *Making: Anthropology, Archaeology, Art and Architecture.* London and New York: Routledge.

Larkin, B. (2013) The Politics and Poetics of Infrastructure. *The Annual Review of Anthropology* 42, 327–343.

Latour, B. (1999) *Pandora's Hope: Essays on the Reality of Science Studies.* Boston, MA: Harvard University Press.

Oliveros, P. (1996) *Four Meditations for Orchestra.* Kingston, NY: Deep Listening Publications.

Oliveros, P. (1971) *Sonic Mediations* [musical score]. Baltimore, MD: Smith Publications American Music.

Özkul, D. (2015) Mobile Communication Technologies and Spatial Perception: Mapping London. In R. Wilken and G. Goggin (eds.), *Locative Media*. New York: Routledge, pp. 39–51.

Shaw, T. and Bowers, J. (2015) Public Making: Artistic Strategies for Working with Museum Collections, Technologies and Publics. In ISEA Vancouver.

Shaw, T. and Bowers, J. (2020) Ambulation: Exploring Listening Technologies for an Extended Sound Walking Practice. *Proceedings of New Interfaces for Musical Expression*.

Shaw, T., Piquemal, S., and Bowers, J. (2015) Fields – An Exploration into the Use of Mobile Devices as a Medium for Sound Diffusion. *Proceedings of New Interfaces for Musical Expression*.

Stankievech, C. (2007) Headphones, Epoche, and *L'extimité*: A Phenomenology of Interiority. *OFFSCREEN* 11(8–9).

Westerkamp, H. (1974) Soundwalking. *Sound Heritage* 3(4).

Westerkamp, H. (2015) The Disruptive Nature of Listening (keynote at ISEA2015), Keynote Address International Symposium on Electronic Art, Vancouver, BC. Canada August 18, 2015. https://www.hildegardwesterkamp.ca/writings/writingsby/?post_id=11&title=the-disruptive-nature-of-listening

Wilken, R. and Goggin, G. (2014) *Locative Media*. London and New York: Routledge.

8

SOUNDWALKING ON THE EDGES: LAND, SAFETY, AND PRIVILEGE IN SÃO PAULO

Paola Cossermelli Messina

The city of São Paulo and the region where it is located are on the ancestral territories and unceded lands of the Mbya, Guarani, and Puri. This acknowledgement is written as violence against Indigenous peoples across Brazil continues. The organization *Articulação dos Povos Indígenas do Brasil* (APIB) is mobilizing regional and local actions in their struggle for sovereignty. For more information, visit apiboficial.org and expand the network for solidarity.

Pitanguá

Though the city of São Paulo is sprawling and diversely inhabited, its soundscape is predominantly peppered with the sounds of two airborne surveyors, airplanes and *bem-te-vi*. The latter are songbirds whose Portuguese name is an onomatopoeia of the sound of their call, which translates to: 'I see you.' Within the Tupi-Guarani linguistic family,[1] this bird is known as the *pitanguá*, meaning the eater of *pitanga*, a small, bright, orange-reddish botanical berry that is native to the Atlantic Forest. In the act of naming this bird, the colonizer essentially transliterates the sound of birdsong. In Tupi-Guarani, the bird's referentiality is construed through its relation to the Land. The act of listening is 'always a listening through, or in relation with land' (Robinson, 2020, p. 53). A soundwalk compounds that sensorial relation to Land in its 'grounded' enactment. Soundwalking can be simultaneously a walking *and* listening through or in relation with Land. Though soundwalking has been historically informed by its relation to 'place', thinking about a relation to Land instead could lead the practice towards what Dylan Robinson describes[2] as 'stratified positional listening toward intersectional antiracist, decolonial, queer, and feminist listening practices' (Robinson, 2020, p. 60).

Sandra Styres capitalizes the word 'Land' as an Indigenous philosophical construct and decolonizing praxis, which indicates 'both space (abstract) and place/land

DOI: 10.4324/9781003193135-9

(concrete)' as 'conceptual, experiential, relational, and embodied' (Styres, 2018, p. 2). Though Styres writes of the potential for such a construct in pedagogical settings, the 'storied landscape' she describes is not only where we walk and listen, but how and why we walk and listen; while Hildegard Westerkamp (2001) describes a soundwalk as 'any excursion whose main purpose is listening to the environment' and specifically focuses on its potential as a practice to retune our (urban) ears to the sounds of nature. The use of the word 'environment' by Westerkamp references the physical geographic space that is part of Land, while excluding its 'principles, philosophies, and ontologies' (Styres, 2018, p. 27). In an extreme urban environment like São Paulo, where the sounds of airplanes roll in and away like the waves on which Portuguese colonizers arrived, a soundwalk is inextricable as an embodied experience of Land, raising questions and significances that can go beyond the dichotomy of urban/rural forms of listening.

Today, the Indigenous storied landscape – 'the spatial and temporal tracks left by our ancestors that can be *read*' (Styres, 2018, p. 28; emphasis hers) – of São Paulo's capital is often not as visibly apparent as it is audible. Monuments, street names, and institutions commemorate the colonizing mission of the Portuguese and the contributions of immigrants who arrived in succession. Directly in front of *Parque Ibirapuera* – which was modelled after the parks of cities like New York – is a statue in tribute to the Portuguese *Bandeirantes* who enslaved and massacred Indigenous and African slaves, a visual reminder of the city's history of settler-colonial violence and its continued glorification. Yet the name Ibirapuera is a word in Tupi-Guarani which means 'rotting tree'. When vocalized, this word is a sonic reference to the village which used to be there and the floods that occurred in that region. However, on its official page, the park pays tribute to a governmental worker, Manuel Lopes de Oliveira, who successfully planted trees in the area to stop the flooding and excessive humidity.[3]

A soundwalk can be a tool for peeling back layers of concrete which, at first listen, appear to sanitize the urban environment of stories which do not exalt its oppressive foundations. While it has more commonly served apolitical purposes and been concerned with the immediate, aesthetic dimensions of traversed environments, I believe soundwalking could be one iteration of what Dylan Robinson describes as a 'decolonial practice of critical listening positionality' which 'actively seeks out (or allows itself) to become haunted' (Robinson, 2020, p. 62). By walking and listening through layers of sound, history, and intersecting positionalities – in this case, specifically as *paulistanos*[4] in positions of privilege who are often discouraged or disconnected from doing so – I contend that it is possible to 'become haunted'. As such, soundwalking while centring stories of Land may serve as a foundation for solidarity and action.

The nuanced observations produced from these soundwalks can also serve to break with broad national perspectives and culturalist explanations which characterize Brazil as particularly 'noisy' in comparison with other countries.[5] Similar internal dichotomies, such as the contrasting of the cities of Rio de Janeiro and São Paulo (and by extension, *cariocas*[6] and *paulistanos*) around levels of violence and capitalist potential, may also be contended.

Listening to the knife sharpener

When visiting family in São Paulo, the city where I was born and lived until my college years, I rarely walk. Depending on where one needs to go and when one needs to be there, walking may not be a feasible option. The capital of the state of São Paulo is infamously plagued by mobility issues and, more often than not, a private car is touted as the most efficient way of getting around.[7] Other factors which contribute to the opting out of public transportation include the limited reach of the subway system, overcrowded buses and trains, sexual harassment on public transport often going unpunished, and price hikes that lead to no infrastructural improvements. Some authors suggest that the June 2013 protests in Brazil – part of the largest country-wide demonstration in two decades – were set off by actions of the *Movimento Passe Livre* ('Free Fare Movement') against increased ticket prices for buses and trains, and later encompassed additional and sometimes contradictory concerns such as corruption in the political class, LGBTQ discrimination, police brutality, and even calls for a return to a military regime (Jiménez-Martínez, 2020, p. 5).

In coming to terms with the fact that my sensory experience of the city had and, in many ways, continues to be mediated by fear, segregation, and vigilance, in late 2018 I decided to take soundwalks in parts of São Paulo which I had often only traversed inside a car. In 'The Sweet Sounds of Havana: Space, Listening, and the Making of Sonic Citizenship', Vincent Adrisani talks about listening to sounds which move across built environments in everyday moments – such as the sound of ice cream vendors in Havana – as generative of a 'sonic citizenship'. He describes this concept as a claiming of space 'according to the aims, aspirations, and desires of those who live there' (Adrisani, 2015). Though Adrisani is reclaiming the concept of citizenship from the grasps of its institutionalized status of nationalist and classist exclusion – imbuing it with a sense of community building and belonging that is not imposed or recognized from the top down – it is difficult to erase the possibility of 'non-citizenship' that comes with it. If the ice cream vendors and other such local actors in Havana without access to political power are generating sonic citizenship, who are the sound producers with the opposite or no such citizenship forming effect?

It is also important to recognize the romanticization of street vendor sounds, particularly by nostalgic ears that recall them from a future space and/or time in which they are no longer heard. In the early 1990s in São Paulo, a new city law prohibited honking by cooking gas supplier trucks.[8] As such, their new iconic sound became a repeating sample of Beethoven's well-known piece for solo piano *Bagatelle No. 25 in A Minor* (more commonly known as 'Für Elise'). Though there were many other royalty-free contenders, the final choice of a classical European piece is worth examining – particularly in considering the power and privilege of the ears that would have found that melody appealing.

The high-pitched whistle of the knife sharpener, however, stands out in my sonic memory most of all.[9] Listening to that call coming down my street, from my place on the thirteenth floor of a gated building in the affluential *Jardins* neighbourhood in São Paulo, had a distorting effect on the sound itself and the possibility of a resulting sonic citizenship. The simple melody of the knife reverberates off cement surfaces and the tall buildings which line the streets of this and many other neighbourhoods in São Paulo, making it difficult to discern how physically close they are at any given time. The indiscernibility of this separation between the listener and the producer of the sound, rendered sonically, is emblematic of class differences – one among many factors which complicate Adrisani's sense of sonic citizenship.

It would be a stretch and falsity to propose that soundwalking as a method can alone be a tool to 'even the playing field' by reducing the aforementioned distance between listener and producer of sound. However, it does allow for sounds of an urban environment as compartmentalized as São Paulo's – cars may have additional layers of insulation from the outside world such as dark tinted windows and bulletproof treatments – to be experienced more directly and thus, potentially embodied by those who are often (by choice, consequence, or both) shielded from such stimuli. In São Paulo, an inter-sensory engagement with my surroundings was often limited in time and scope, as I moved cautiously between interiors – private vehicles, schools, 30-storey buildings towering high above the streets, and shopping centres patrolled by armed guards. At night, as I laid down to sleep, the groans of rusty trucks straining and motorcycle exhaust pipes blasting echoed in lively conversation with each other as they climbed up the steep slope of my street.

The potential for a soundwalk in my home city of São Paulo – as an intersensorial exercise which enables 'meaningful perceptual and embodied experiences of our places' (Styres, 2018, p. 25) – comes through when it is directed towards an examination of intersecting positionalities in relation to Land. Being of mixed Southwest Asian and European descent in Brazil is an ethnic identifier of whiteness and, more often than not, an indicator of a comfortable living situation. My school uniform, with its red blazer and dark grey skirt, the uncommonly green neighbourhood where most of my daily routines took place, and the double-gated apartment building I lived in, were all indicators of this. The very fact that I was not encouraged to walk – and by extension, truly listen to – the city that I lived in, as a measure of safety, is perhaps the most telling of all.

Conducting these soundwalks was especially significant as I've lived most of the past 15 years of my life as an immigrant in the United States. An unstable and disorienting feeling lives in my body, as I inhabit places dichotomized as 'home' and 'away', framed by citizenship statuses, my queerness and its acceptability, etc. Striving towards orientation, 'aligning body and space' (Ahmed, 2006, p. 7) – in other words, 'citizenship' – is not going to resolve that disorientation or produce any meaningful outcome. To conduct a soundwalk without the goal of orienting oneself, but to explore that disorientation beyond abstract concepts of 'space' and 'place' is to allow 'the oblique to open up another angle on the world' (Ahmed 2006, p. 172).

Soundwalks from Rua Juquiá and Avenida Paulista

At the end of 2018, I recorded soundwalks on two routes (Figure 8.1), which were only loosely planned, in hopes of engaging with some of my personal fears

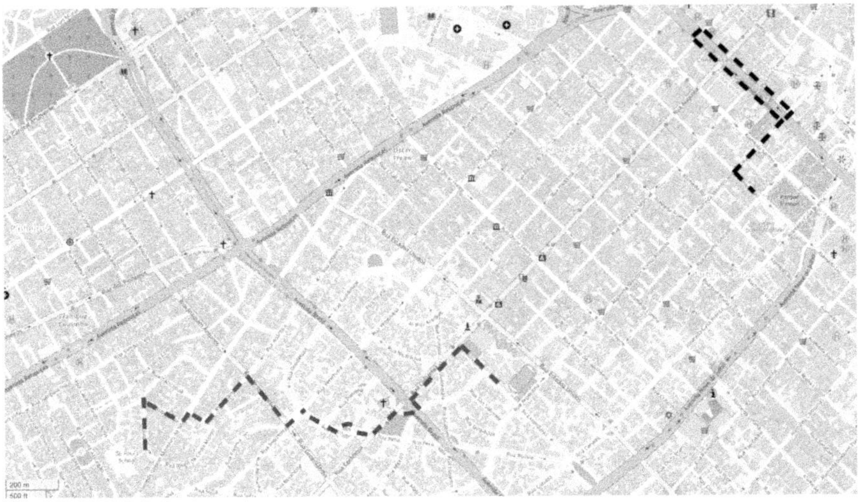

FIGURE 8.1 A map depicting the author's two soundwalks in São Paulo. The lower dashed path begins at the marker on Rua Juquiá and ends slightly past Avenida Brasil. The upper dashed path indicates a soundwalk along part of Avenida Paulista. Map data © OpenStreetMap and contributors, CC-BY-SA (openstreetmap.org/copyright)

and fluctuating positionalities as a queer, white/person of colour, expatriate/immigrant. I walked with an awareness of the microcosm of São Paulo which I was experiencing, the storied landscape on which I was walking and listening, and a heightened attention to the sounds which had enveloped my day to day for many years.

The first walk I will describe was a morning walk between my former high school (Figure 8.2) and home in the *Zona Oeste* of São Paulo, while the second was an early evening walk on *Avenida Paulista*. Often just called *Paulista*, this is a bustling, iconic avenue at the top of one of the steepest hills in the city, located at the crux of the Western, Central, and Southern regions of São Paulo. Where the first soundwalk ends is merely a ten-minute walk from where the other begins. These are physically close but sonically contrasting areas, both of which are significant to my personal experience growing up in São Paulo, but were never experienced quite in this way before. In the final section of this chapter, I will describe both of these soundwalks as they relate to the aforementioned discussions around sonic citizenship and make a case for the potential of soundwalking as a methodology when it is specifically thought of in relation to Land.

Juquiá

Juquiá is a word in Tupi-Guarani which could mean 'dirty river' or a trap to capture fish. In São Paulo, the street named *Juquiá* is lined with trees and walled-off houses. With no river or source of water in sight, today it could be a testament to what Eduardo de Almeida Navarro dubs *toponímia indígena artificial* ('artificial indigenous toponym') in Brazil. Navarro classifies place names as either: spontaneous, which are assigned by speakers of the language to physical and natural features such as waterfalls, hills, mountains; or artificial, assigned by official initiatives to streets, monuments, municipalities, etc. (Navarro, 2020, p. 255). The intention behind the use of an artificial toponym depends on the context, but Navarro attributes the specific case of indigenous languages in Brazil to a mythologization and homogenization of the diverse indigenous nations on that Land to a singular Tupi group and language (the Tupiniquins, Tupinambás, Potiguaras, and several others were largely ignored).

It could be that a river once existed in this place, but in my childhood memory, Rua Juquiá was filled to the brim with cars as early as seven in the morning. I attended a private British school for kindergarteners through to high school students. It was and continues to be the only non-residential building on Rua Juquiá. During the day, students' personal security guards and drivers would find a place to park and nap until the final school bell rang. I recall making beelines from my mother's car straight to the school gate, motivated by tales of *sequestros relâmpagos*, which translates to 'lightning kidnappings'. Young people going to private schools were said to be the primary targets of these sorts of kidnappings and were held for ransom. That short distance between safe interiors sometimes made its way into my dreams. Once, likely in a state of sleep deprivation, I got my thumb caught in the car door as I was racing out.

FIGURE 8.2 On Rua Juquiá, the façade of the author's high school is barely visible over a massive wall. A security camera sits on top of it, overlooking the sidewalk below

FIGURE 8.3 A group of cameras are mounted on a pole no more than two metres high beside a tree on Rua Jacupiranga (perpendicular to Rua Juquiá). Below them, a yellow sign reads 'City Cameras' in English

On the morning of 29 December 2018, the hum of car engines and sounds of children racing to make it to class on time were completely absent. There were only a few parked cars and hardly any people on the street. I looked up at the wall obstructing my school from view (Figure 8.2). The walls were initially put up at some point during my senior year in 2005, but have been given added height and bulbous black-eyed cameras to scan the streets from every angle. On Rua Jacupiranga (a Tupi-Guarani word meaning red *jacu*, a kind of bird that eats grains) perpendicular to Rua Juquiá, there is a new addition – a set of so-called 'City Cameras', curiously placed at eye level (Figure 8.3). This is hardly what journalist Jane Jacobs meant when she argued for the importance of 'eyes on the street'. Buildings which are oriented towards the street, she says, contribute towards the safety of residents and passers-by.[10] If the buildings are walled off and the 'eyes' are those of surveillance cameras, a sense of danger and unwelcomeness is only heightened. Significantly, the impact that this visual surveillance has on the sonic activity of the street is drastic. The sounds and movement which fill Rua Juquiá on a regular school day are deceptive and not representative of its actual sonic production. The looming walls of a private institution and millionaire real estate, with their accompanying surveillance cameras, are producers of silence. This is not to say that the recording of my soundwalk is actually sound-less but that silence is present as a 'deployment of the limit' (Ochoa Gautier, 2015, p. 184) of sonic production, imposed by property owners of Rua Juquiá as a boundary to participation and the city's collective potential.[11] The 'sinister dimension' (ibid., p. 183) of this particular kind of silence echoed especially true in the wake of Jair Bolsonaro's successful election a little more than a month before this walk, in November 2018. The Brazilian myth of racial democracy, buttressed by Gilberto Freyre's valorization of *mestiçagem* ('miscegenation'), was notably countered in the years that the *Partido dos Trabalhadores* (PT) held the presidency, yet the increase in institutional policies against racism hardly put a dent in incarceration and homicide rates among Afro-Brazilians. With Bolsonaro at the helm, the silent segregation of streets such as these stands in stark contrast to the voiced, brazen racism, such as his comparison of black *Quilombola* people to cattle and a slew of other remarks throughout his presidency (Alfonso, 2019).

Ultimately, the silence that surveillance imposes on Rua Juquiá, and the outnumbering of human and more-than-human presences by vehicles, warrants a return to the question: who are the 'sonic citizens' of this street? (Figure 8.4). Other than me, only a few construction workers shovelling bits of cement into a bin and security guards standing outside of houses (additional 'eyes on the street') were present. Employment and services offered give non-residents temporary passes to be on streets like Rua Juquiá. The security guards watched me for a brief moment – an unfamiliar presence walking on a street which is often traversed by cars – but concluded soon enough that I was no threat to the houses they were employed to protect.

The songs of several *bem-te-vi*, flitting about in the green canopies above, are still a constant sonic presence. The surrounding neighbourhood where Rua Juquiá is located, known as *Jardins* ('gardens'), is one of the greenest in the city, yet only birds voice their presence and delight in this São Paulo rarity. I can't help but think of them as the true sonic citizens of Rua Juquiá. If sonic citizenship is about the less powerful establishing their presence and belonging in public spaces through sound, the concept rings true in a different way in the case of Rua Juquiá. While human presence is questioned and restricted in many ways (evidencing the exclusionary side of a sonic citizenship), the *bem-te-vi* has adapted to the urban environment with both its flexible eating habits and ability to insert itself sonically, maintaining its birdsong audible despite the

FIGURE 8.4 Looking up Rua Juquiá on 29 December 2018. A few trees line the street and most houses are obscured by tall walls. Only a couple of security guards can be seen standing outside, near a white guard hut

intrusiveness and loudness of the urban soundscape. If the category of citizenship should be attributed at all is a question I address in the final section of this chapter.

As I ventured further away from Rua Juquiá, towards Avenida Brasil – one of the main arteries of the Jardins neighbourhood – the baseline hum of São Paulo's traffic reoccupied the background layer of my recording. The green canopy which envelops the residential streets has become sparser, though every now and then a corner boasts a giant rubber tree (*falsa-seringueira*) or two (Figure 8.5). These are named *falsa* ('fake') because they were brought to Brazil from India and southeast Asia via colonization, masking the years of violence that have been inflicted upon the *seringueira* that is native to Brazil and those who live in relation to it.[12]

FIGURE 8.5 A pair of *falsa-seringueiras* ('fake' rubber trees) near Avenida Brasil

Up until this point, I was carrying my equipment – a Zoom H6 and Rode NTG 2 shotgun microphone – inside a bag, in order to draw less attention to myself. This is certainly a decision informed by the same fear that would make me hurry from the car to the school gate. As a consequence of this, throughout most of the recordings here (and in the next section, *Paulista*), one can hear a light, rhythmic thudding from the microphone hitting the inside of my tote bag. Consider this another layer of distortion to the sound of my environment, like the vertical distance to the call of the knife sharpener. I produce a sonic manifestation of the danger and vigilance I have been told to feel.

Paulista

Only a 20-minute walk away from Rua Juquiá is *Avenida Paulista*, an almost 3 km-long avenue with ample sidewalks, designed by Joaquim Eugênio de Lima (1845–1902) in 1891 with the boulevards of Europe and North America in mind. The luxurious mansions which first lined the avenue were owned by coffee plantation owners and immigrant entrepreneurial families. In 1909, it became the first paved road in the city. Today, its elevated position in São Paulo acts as a dividing line between the capital's different 'faces' and temporal realities– the *centro* ('centre') and *Zona Oeste* (the 'Western' region). When I was a teenager, I knew that if I turned off Avenida Paulista onto *Rua Augusta* towards *Baixo Augusta* ('low' Augusta) and the city centre, I would find queer-friendly bars and clubs. I remember driving by them with my parents; it was implied that there was danger there, too, though stories of violence (like the 'lightning kidnappings') were not available, the prostitution and 'deviant' sexual activities of the area were enough to signal a threat. If I took Augusta down the other side of the hill, I would be in one of the most affluent neighbourhoods in the city.

While Avenida Paulista, also known simply as Paulista, was designed for mansion owners and continues to be a hub for office workers, from the 1990s on it became one of the main stages for protests and political demonstrations. Nearby lies the heart of the city centre, a so-called 'urban void', which has been the site of vacant building occupations, homelessness, and drug-use.

To practise soundwalking on Avenida Paulista, where the Portuguese colonial project bore fruit in an urban form based on Western aesthetics and norms, is to imagine a foundation built by the hands of West African and Indigenous peoples. On the night I recorded this soundwalk, the air was buzzing with chatter, a mélange of street musicians and performers, and skateboard decks slamming down and grating on cement. The abundance of human sounds stood in stark contrast to Rua Juquiá, as did the sheer number and variety of them; teenagers conversing excitedly while on holiday from school, people in business suits making beelines to the subway with their phones glued to their ears, motorcycles beeping to warn drivers of their lane-weaving tendencies, and the constant arrivals and departures of two-section city busses that squeak to a stop and hydraulics that seem to lower them to the curb with a sigh. My recording captured the sounds of many performances as well, including one of a guitar player in the reverberant entry to a mall, singing a Caetano Veloso song from the 1980s, *Você é Linda* ('You Are Beautiful'). The sound of traffic is as ubiquitous as that of the *bem-te-vi*, which are inaudible within the many layered soundscape of the avenue, but undoubtedly remain in the vicinity. Generally, the presence of the city police is widespread and visible on Avenida Paulista, though they tend to spend most of their time as silent observers. Their 'sinister' and limiting potential is akin to the surveillance cameras on Rua Juquiá. Sirens may suddenly blare and echo down the avenue as an audible, shocking reminder of the futility of 'sonic citizenship' in the face of such violent authority.

A soundwalk on Avenida Paulista gives the impression of endless possibility to engage in what could be called a sonic citizenship, but I am more inclined to use a word like 'belonging'. The main point I would like to make, however, is that the belonging I felt is reserved for my own ears – walking and listening in the privilege I hold. While on Rua Juquiá, this possibility remained complicated by fear, vigilance, and a vacuum of human activity. As a place where leisure, protest, and capitalism thrive under one roof, Avenida Paulista is presented as a beacon of vocalization and making oneself heard, but its historical foundation and current police presence tells us something more. Questioning the feeling of sonic belonging is essential and possible while soundwalking in relation to Land.

Nhamonhendu

Nhamonhendu is an expression often found in many *mborai* ('songs') of the Guarani-Mbya people. It roughly translates as 'to make sound', but with the *-monhendu* suffix simultaneously implies 'to make one listen to' or to 'making oneself heard'.[13] In other words, to produce sound is to be heard/listened to (Andriolo Abel, 2019, p. 44) – a co-existence which is not a part of the lexicon of colonialism.[14] In fact, the English language further complicates the prospect of 'making oneself heard' by distinguishing between 'to listen' and 'to hear'; while the latter implies a more attentive form of receiving a sound, the former is more passive.

The practice of soundwalking in relation to Land opens up the possibility for this necessary articulation. Rather than framing participation and collective potential under the guise of concepts like citizenship – which romanticize urban contexts, nation-building, and ultimately imply power *with* disempowerment – I would suggest a questioning around the notion of belonging in itself. Soundwalking as a practice has the potential – in its 'grounded' enactment of listening – to recognize and reflect on past and current exclusions which come at the cost of one's own sense of belonging. In other words, a walking *and* listening through or in relation with Land. As such, a soundwalk's potential is contingent on the soundwalker's relation to Land as more than a 'physical geographic space'. Rather, Land as an expression of:

> a duality that refers not only to place as a physical geographic space but also to the underlying conceptual principles, philosophies, and ontologies of that space […] Land, as a thermotical and philosophical concept, comprises storied and journeyed connections of self-in-relationship – to each other, to our places, and to all of creation – as a central model for interpretation and meaning-making.
>
> *(Styres, 2018, pp. 27–28)*

To conduct a soundwalk reflecting upon the immense scope of what Land encompasses – and how our own sense of belonging affects the walking, listening, and sounding of others – is to go beyond simply *understanding* who we are in relation to Land, but to reinvent ourselves in and through that relation.

Notes

1 Tupi-Guarani is a branch of the Tupi language family which includes Tupinamba, Guarani, and a number of other languages spoken from French Guiana south to Paraguay and west to eastern Colombia.

2 Dylan Robinson (2020) expands upon Martin Daughtry's work on forms of critical listening elaborated in the article 'Acoustic Palimpsests and the Politics of Listening' (2013).
3 Parque Ibirapuera Conservação (2022) *História e Fotografias Históricas*. Available at: https://parqueibirapuera.org/parque-ibirapuera/historia-mais-completa/ (accessed 18 July 2022).
4 *Paulistanos* is the Portuguese word for those who live in the city of São Paulo, while *paulistas* refers to inhabitants of the state of São Paulo.
5 Cardoso, L. (2019) *Sound-Politics in São Paulo*. Oxford: Oxford University Press.
6 *Cariocas* is the Portuguese word for those who live in the city of Rio de Janeiro, though the original meaning of the word is indigenous and debated.
7 Rolnik, R., and Klintowitz, D. (2011) '(Im)Mobility in the city of São Paulo,' *Dossier São Paulo*, Av. 25(71).
8 Larissa Linder, Edison Veiga. *Como Beethoven virou até 'musiquinha do gás' no Brasil*. Available at: https://www.dw.com/pt-br/como-beethoven-virou-at%C3%A9-musiquinha-do-g%C3%A1s-no-brasil/a-55971420 (accessed 20 July 2022).
9 Carlos Bozzo Junior (2019). *Sons de S. Paulo: O Amolador* [online video]. Available at: https://www.youtube.com/watch?v=AizMxcO3GEY (accessed 18 July 2022).
10 In her book *The Death and Life of Great American Cities* (1961), Jane Jacobs argues that high activity and intermingling on city streets contribute towards a sense of security and vibrancy.
11 In other words, silence is not just the absence of sound, but a limitation to the potential of sound production. In this case, enforced by the residents of Rua Juquiá through surveillance.
12 A violent history of extractivism in the Brazilian Amazon – which includes massive deforestation, genocide of multiple indigenous groups, and slavery – has centred on the commercialization of the *seringueira*, Brazil's native rubber tree. For more on this topic, see Violeta Refkalefsky Loureiro's (2002) Amazônia: uma história de perdas e danos, um futuro a (re)construer. *Estudos Avançados* 16(45).
13 Valéria Macedo explains that the performance of *mborai* (songs, also called *poraí*) by the Guarani often feature the expression *nhamonhandu* in reference to *Nhanderu* and *Nhandexy kuéry* (literally 'Our Fathers' and 'Our Mothers', immortals who inhabit a cosmic domain), who are 'both the source of songs and to whom they are commonly addressed' (Macedo, 2011, p. 405).
14 In fact, the Guarani say that the very presence of jurua (non-indigenous people) during shamanic chants hinders the possibility for nhamonhendu. Youth choirs present 'Guarani culture' through a particular form of mborai that is not restricted in its circulation and shows a 'considerable investment … to making themselves heard, perceived (and respected by whites through songs)' (Macedo, 2011, p. 405).

References

Adrisani, V. (2015) The Sweet Sounds of Havana: Space, Listening, and the Making of Sonic Citizenship. Available at: https://soundstudiesblog.com/2015/09/17/the-sweet-sounds-of-havana-space-listening-and-the-making-of-sonic-citizenship/ (accessed 18 July 2022).

Ahmed, S. (2006) *Queer Phenomenology: Orientations, Objects, Others*. Durham: Duke University Press.

Alfonso, D.A. (2019) Bolsonaro's Take on the 'Absence of Racism' in Brazil. *Institute of Race Relations*, 61(3), 33–49.

Andriolo Abel, R. (2019) '*Lá no Alto se Canta o Tempo Inteiro': Formas de ensinamentos Guarani-Mbya e o potencial do canto como (trans)formação*. Dissertation, Universidade Federal de Santa Catarina, Florianópolis.

Cardoso, L. (2019) *Sound-Politics in São Paulo*. Oxford: Oxford University Press.
Daughtry, J.M. (2013) Acoustic Palimpsests and the Politics of Listening. *Music & Politics* 7(1).
Freyre, G. (1946) *Casa Grande e Senzala* (translated into English as *The Masters and the Slaves* [1986]. Oakland, CA: University of California Press).
Jacobs, J. (1961) *The Death and Life of Great American Cities*. New York: Random House.
Jiménez-Martínez, C. (2020) Introduction: The June 2013 Protests and the Image of Brazil. In: *Media and the Image of the Nation during Brazil's 2013 Protests*. Cham: Palgrave Macmillan.
Macedo, V. (2011) Tracking Guarani Songs: Between Villages, Cities and Worlds. *Vibrant: Virtual Brazilian Anthropology* 8(1), 377–410.
Navarro, E. de Almeida. (2020) Artificial Indigenous Place Names in Brazil: A Classification of Tupi Origin Names Created in the 19th and 20th Centuries. In: Revista Letras Raras. *Campina Grande* 9(2), 252–267.
Ochoa-Gautier, A.M. (2015) Silence. In: D. Novak and M. Sakakeeny, eds. *Keywords in Sound*. Durham: Duke University Press, pp. 183–192.
Refkalefsky Loureiro, V. (2002) Amazônia: uma história de perdas e danos, um futuro a (re) construir. *Estudos Avançados* 16(45).
Robinson, D. (2020) *Hungry Listening*. Minneapolis: University of Minnesota Press.
Rolnik, R. and Klintowitz, D. (2011) (Im)Mobility in the city of São Paulo. *Dossier São Paulo*, Av. 25(71).
Styres, S. (2018) Literacies of Land: Decolonizing Narratives, Storying, and Literature. In: E. Tuck, L. Tuhiwai Smith, and K. Wayne Yang, eds. *Indigenous and Decolonizing Studies in Education: Mapping the Long View*. New York: Routledge, pp. 24–37.
Westerkamp, H. (2001) Soundwalking. In: A. Carlyle, ed. *Autumn Leaves, Sound and the Environment in Artistic Practice*. Paris: Double Entendre.

9

TRAVERSING THE CONCRETE MARVEL: WALKING THE UN-SUNG CITY

Katt Hernandez

In this chapter I will take up my own way of walking, which could be described as lying askew from psychogeographical *dérive* (or 'drifting') and 'soundwalking'. My walking is wholly situated in cities and urban suburbs – the only environment I have lived in. But the original premises of soundwalking can be lacking when it comes to addressing people who dwell in or relate to heavily human-built environments like urban or industrial areas, or exploited land. And while psychogeography is closer to my own wandering, it also emerges from a very specific context, different from my own. My experience of walking necessitates the creation of both alternatives to and development past some well-established premises within the origins of psychogeography and soundwalking, because it leads me to traverse places, encounter people and listen to things in them that do not fit their original agendas in either. Here, then, I will wander through some prior walks, sharing some of the places I have traversed that illuminate this.

I began soundwalking long before I knew that word, or others that could articulate something about the walks I made, like psychogeography or *dérive*. But I began, all the same, both to drift and to listen, and occasionally to play back to or with the places I traversed on my violin, as a teenager. I was drawn through lunch counters and late-night coffee houses, pre-laptop, to converse with the chess-playing poets, war veterans, self-made artists and anti-war activists who populated the edges of the university in my hometown of Ann Arbor. I wandered the old cemetery, greeting familiar graves, then walked the calm, white corridors of the hospital, stopping sometimes to talk with patients or use the payphone. I climbed up on rooftops and fire escapes, or walked off the paths in parks, into the topiary gardens of the lavish houses at its borders, or down the diminutive Huron river as far as I thought I could go. I wandered the aluminium-sided suburbs, big-box stores and strip-mall stretches of main outer byways. In those days, pre-keycard, I wandered inside the university, sneaking past the guards to explore the endless

DOI: 10.4324/9781003193135-10

rooms and corridors, or glean through massive libraries. I walked the streets of the little city and its outskirts at every possible hour of day or night, conversing with the ghosts, birds, passers-by and buildings as I went.

Moving to the east coast brought a vast expansion of this pursuit. New environments – beautiful or ugly, sketchy or lively, huge or intimate – rolled out over seemingly infinite streets from Boston to Baltimore. Walking went from cycling though the familiar to tromping deeper into re-charted territory, usually with the violin by my side. From the small town to the large city, the sounds of traffic calmed my dreams like the ocean, and the sound of conversation raised high over the din gave me to think, laugh or join. The tumble of different music, pouring out of shops, cars and industry speakers, became an oracular beacon; mechanical and human spinning alike marked out the vast cityscape of the Bos-Wash corridor in ever wider paths as I passed.

I learned the word 'gentrification' when I was 23, and felt deep relief at having it to describe what I had previously called 'the nothing', taken from *The Neverending Story* (Ende, 1979); the silenc(ing), inexorable force that drained nearly all life and possibility out of my first hometown and my second. I am wholly based in the urban environment, with all its injustices and troubles knotted inextricably through the fabric of its marvels and good stories, wondrous and tragic alike. So this force called gentrification has always held the lion's share of sway over my most subjective sense of injustice and destruction, even though there are, objectively, far greater ills in the world. Meanwhile, I did not learn the words psychogeography, *dérive* or soundwalk until I was in my forties and living in Sweden. These terms of articulation for practices that are so vital to me both flummox the fundamental spontaneity and outsiderness that speak to me in them, and imbue them with deep new facets and possibilities.

The origins of the soundwalk

Like many, I first encountered the term 'soundwalk' reading R. Murray Schafer's work, from which great numbers of musicians, sound artists, environmentalists and others have taken inspiration. But I never sat comfortably with his definitions of high and low 'fidelity' sound, nor his demand for 'ear cleaning', which consequently demands the 'crack-up' of most of the sounds that inform my life-long existence as a city-dweller. So I never engaged fully with the subsequent lines of practice that issued from his work, and sought kinship elsewhere to articulate what it is I do when I walk and listen (Schafer, 1977, p. 164). While the shifting sheets of sound in urban environments certainly do contain fistfuls of oppressions from noise pollution, class divisions and multiple segregations, they are what I know as home. The idea that children who grow up in cities must already be cleaning their ears of their own primordial environment is one much of my own experience lies to the opposite of; I celebrate those sounds in my walks (Schafer, 1977, p. 173). Cities are full of still more contradictory bubbles: housing bubbles that exclude all but the wealthy; institutional bubbles, alternately opening doors wide and slamming them shut tight to street and community alike; bubbles of locally owned businesses, complete with their own internal hierarchies which might be kind or dreadful, split apart by bubbles of franchises; bubbles of privilege and oppression; bubbles of resistance and survival – all pushing against one another in uneven force. Traversing them on foot as far as their boundaries allow, and transversing those boundaries where it is possible is a way to refute those hierarchies. A tiny, not particularly organized or effective way. But a way.

FIGURE 9.1 Old as Hills. Photo: Katt Hernandez, 2013, all rights reserved

The serene summer sight of the Goodyear blimp over Ann Arbor; the 'Liquor – Old as Hills' shop at the top of the Delancey street station stairs in New York; the flower gardens grown on small bits of lawn in cinderblocks, lovingly watched over by the Virgin Mary in up-turned bathtubs in Somerville – all these are part of the All-Fidelity world in which I dwell (Figure 9.1).

The closest practice I have found to my way of soundwalking does not come from the literature where the term was coined at all, but from psychogeography: a vision of transforming place through internal and external re-routing devised by the Situationists in 1950s Paris (Debord, 1955 and Chtcheglov, 1953). The primary practice of psychogeography is the *dérive*, which means 'drifting' or 'aimless wandering', taking in *all* one perceives as newly arranged information, in order to expand the perception of place as a radical and creative act. Ultimately, listening is utterly subjective, and so is the *dérive*. Here, the act of listening can be of higher or lower quality, but the quality of the *focal point* of that listening is arbitrary by default.

A beautiful noise

FIGURE 9.2 Västberga Red Bridge Winter. Photo: Katt Hernandez, 2020. All rights reserved

The composer Toru Takemitsu described the buzzing of the string of an instrument called the *biwa* against its face-plate as 'a beautiful noise' (Takemitsu, 1995, p. 64). I first read this in the 1990s, as I searched for new ways of playing the violin, and sought the same 'beautiful noise' in extended techniques. I began to walk the city with my instrument, stopping to play with sounds as if they were fellow improvisors; bridges, ventilation units, boat engines, crowds, and the birds and squirrels so acclimated to urban life (Figure 9.2). In 2015, having returned to institutional electronic music studios after a 17-year hiatus, I found a collection of writings edited by Nilsen and Gandy, entitled *The Acoustic City* (Gandy and Nilsen, 2015). One chapter, by David Novak, took this quote I had been so thunderstruck by years earlier, and expanded it: 'emerging from the apparatus of an obstacle' (Novak, 2015, p. 27). The 'beautiful noise emerging from the apparatus of an obstacle' is nearly always what I seek – the terrible, wonderful spectre of singing in adversity, gathering the daily stories of the cities where I live in footsteps and concerts – the disappearing urban environments I have always felt were my truest home, even as they vanish all about our ears.

Sonic meditations

In 1971, Pauline Oliveros published *Sonic Meditations*, still in heavy use amongst those who engage with the group of practices now known as 'soundwalking' (Oliveros, 1971). Oliveros' practice is more generalized in its hierarchic exclusion of one sound over another than Schafer's. It is less focused on the environment than on specific states of consciousness, and how they 'tune' or transform body, mind and spirit (Oliveros, 1971, p. ii). Her most referenced Sonic Meditation amongst those who engage with soundwalking is: 'Take a walk at night. Walk so silently that the bottoms of your feet become ears' (Oliveros, 1971, p. 6).

Compositional, score-like instructions like these appear throughout *Sonic Meditations*. Like Schafer, Oliveros was still firmly situated a composer, parallel to her more activist, feminist work. The later 'Deep Listening' system for which *Sonic Meditations* is an origin (which Oliveros later trademarked) has an implied disapproval in its very name for sounds that interfere with activities like quiet contemplation and reflection.

But why can't the soles of my feet become ears on the faster-paced streets of a city? Why not in the clomping work shoes I wore as a runner in a paper factory, with its plethora of wonderful machine sounds, or in the grind and crunch of snow-boots on sanded concrete in winter? Couldn't my feet become ears in the multi-faceted daylight? Couldn't machines and post-industrial sounds some deem to be 'noise pollution' be exactly the instruments I want to orchestrate my particular, moving soundscape? Cities are sometimes over-romanticized as archetypes of multi-faceted place, but the way they manifest in daily minutia defies the archetypal. In walking and listening there, the pre-human, artificial, historic, futuristic and contemporary all collide in tangles of pathways and furtive maps.

Engines in serenity

In summer of 2001, I sat by the Boston Harbor, across the water from the airport. I had just walked through the North End. The version of that storied neighbourhood I knew was evaporating with stunning speed. In 1983, my whole idea of what a 'city' is was transformed in an instant by women in house-dresses, hanging laundry out the brick tenement house windows, and standing in make-shift stalls under the highway overpass, fish and tomatoes heavy in the air, cars and trucks roaring overhead. By the late 2000s, it was rapidly transforming into jaw-droppingly expensive condominiums, with city edicts against outdoor laundry and unpermitted fruit stalls – and the highway had been buried beneath a new park. Walking through the eerie quiet, I was comforted by the sight of two elder men sitting on lawn chairs in the middle of the street, as people there had once been wont to do. I continued to the equally quiet piers, where a friend had once waved an arm over the harbour and declared that the soul of Boston died long ago, when it ceased to be a centre of shipping.

On the sunny evening pier, the city faded from sonic view, leaving only the water, the soft wind, the seagulls – and the airplanes. Their engines were distant and serene, looping out over the harbour to land. I listened a long time, wondering at this sonic garden, cloaked in that benign neglect that vanishes in the face of gentrification, urban redevelopment and the cult of the entrepreneur. Two years later, I participated in the Boston Center for the Arts *Boom Box* show, a staggeringly dense affair with masses of sound works placed in exuberant hyper-overstimulation. I wandered into the few separated spaces, seeking rest, and found a room with a 'composition' by John Malia (2003). I write it in quotes because, while I am inspired by much of Malia's work, this was simply a field recording of the very same pier – the water, the gulls and the gliding engines of airplanes careening through the blue sky, defying all they implied in the summer lull of their sound. The solace of listening to them returned in that fleeting imprint. I stood there, torn, between feelings of sonic sisterhood and covetousness over the lost secret – the exploited place I wished had remained un-found, suddenly paired with the word 'composer'. In miles and years of walks since, I haven't return again.

Kingston

In 2005 I was part of an intermedia ensemble that performed at Pauline Oliveros' Deep Listening Space in Kingston, New York.[1] At that time, despairing of ever having access to electroacoustic music facilities again, I had dedicated myself wholly to the violin, morphing into a pantomime of that which I could not have: 'recording' the environment by playing in it, trying to 'transpose' whatever I encountered, and emulating electronic instruments.[2] After the concert, she showed us around: the gallery on the first floor, the small chambers for students on the second, and even a look in at the third, where there was a wonderful music studio. Back downstairs, I found a door to a courtyard with a wooden sculpture, built from heavy blocks with faded red and blue paint, a single, tiny bell hanging on its edge. The more I listened, the more the bell was utterly silent. A part of me wanted to enjoy it. Another part of me recoiled, yearning back to the noisy city and all its mad variety, where similar silences fell like a pall over the windows where laundry was once hung out, the highway spaghetti where trucks once roared over the unlikely humanity below, and the harbours missing the din of ships – neither in animosity nor anachronism, but to the former fabric of everyday life.

Beautiful noise is different for everyone. In Stockholm I came across a room full of men born in the mid-1950s one day, with synthesizers, computers and contact mics. They were trying to recreate the sound of a particular boat engine, imprinted in their memories of an earlier time. The *tändkule* ('hot bulb') engine has a particular noise, spinning with a rich metallic sound, almost as if it were churning some smooth, resonant oil. Its melody can be picked out from the water without needing to be a sailor. It drove smaller boats in the urban waters of Stockholm, a nautical city whose centre is laid out across 14 islands. This group of synthesizer hobbyists tried to recreate the sound, and all the associations it carried, of a foregone Stockholm: a beautiful noise.

I could relate. In years of trips to New York to play concerts, I had spent many late nights on subway platforms, where the 'trash train' – a special subway car equipped to carry off the contents of the garbage cans – was often met with happy, rhetorical greetings by friends and strangers alike. Its motor had a gleefully irreverent crunching sound, simultaneously solid and sounding as if it could break down at any time. Here is where the subjectivity of listening lies: for me, it's beautiful to think of the city's late-night workers, stolidly keeping the impossible subway system up and running. It's beautiful to look at the grimy, solid metalwork of the train; beautiful to listen to those workers, talking in a loud holler; beautiful to listen to the motor wavering between failure and marching on. Even with all the hardships it contains. Even with all the problematic things about petroleum, workers' rights, the environmental impact of living in a city with dirty air. Maybe this thing I find beautiful will have to be replaced one day with something that solves those problems, and maybe I will be lucky enough to find that beautiful, too. In the end, the subjectivity of what one finds beautiful to listen to can't be defined by anyone but the individual doing the listening.

Psychogeography

Although I was a dedicated anti-gentrification activist for almost two decades, I did not come across the term psychogeography until the mid-2010s, after I moved to Sweden. The term's origins lie especially in Guy Debord (1955) and Ivan Chtcheglov (1953), but it has grown to include practitioners both before and after its inception. Here one seeks out new and sometimes jarring ways of traversing places, thus transforming them, by walking algorithmic maps, habit-altering routes and obstacle-filled paths. Psychogeography has since been claimed by everyone from colourfully aliased activists to carefully studied researchers to artists centred upon it.[3]

Some of the contemporary artists who have taken up psychogeography with depth are fairly well known authors. In the early 2000s, Rebecca Solnit wrote movingly in psychogeographical terms of the lost punk and underground scenes of her native San Francisco, the other-worldly landscapes of the west, various sites of sacred acts and activism beyond and the communities and mythologies that surround them (Solnit, 2001; 2006). Will Self's ornery journeys through overlooked and under-sung urban sprawls come even closer to the counter-balance of 'beauty' I seek when I walk the city (Self and Steadman, 2007). And in the opening to his book *The Old Ways*, I find cousinship in Macfarlane's snowy walk through some fairy-like kingdom whose zenith reveals its alter-self as a suburban golf course (Macfarlane, 2013, pp. 2–8).

It is important to point out here that I do not at all consider what I engage in as being related to the act of the *flaneur*. Lauren Elkin understands the historic descriptions of this creature as 'a man about town who comes from sufficient wealth to have the time to wander the city at will, hanging out in cafés, and watching the various inhabitants of the city at work and at play' (2016, p. 21). This is nothing I have ever related to my own way of walking, either as part of my attempt to understand and participate in the lives of the cities where I live, or as part of the music I make. The purposelessness of the *dérive* has a radical, political bent to its non-productive nature; a private striking out for new realms beyond the imprint of the seemingly immutable one, that this 'man about town' is not seeking. Ultimately, the difference between a *flaneur* and a psychogeographer is that the *flaneur* drifts in order to consume, while the psychogeographer drifts in order to affect their own perception in a radical way, and thus refute aspects of the traversed environment. And where the term *flaneur* has the male declension embedded in it, making for the awkwardly gender non-neutral term *flaneuse* as an attempt to refute its embedded sexism, the term psychogeographer is thankfully gender neutral, which gives me the freedom not to define myself by gender within it. In the years of alternating 'temp', shop-keeping, phone room and cafe work I did in my decade of life in Boston, I looked upon my unchosen, unwanted working hours as a kind of *dérive* through the inner commercial, industrial, non-profit and corporate rooms of that city, drifting in the careening wander of unstable employment through all sorts of places, from the unnoticed stance of the secretary, concierge, counter worker or message runner. This turning of wage work into wandering, while simultaneously doing both, is the opposite of *flanerie*.

There are artists in other disciplines who work with more deliberate and multifaceted depth in psychogeography. Geraldine Hudson has worked with psychogeography at the intersection of visual arts and mythologies for many years, with some of her wide-ranging practice rooted in the Bristol punk scene.[4] Amongst the many projects she has made in Stockholm is an exhibition drawn from a psychogeographical walk between the sites of two former mental institutions, or what Hudson terms a 'psychogeomagikal ritual' at the top of an artificially made mountain south of the city, built from the rubble of storied neighbourhoods since torn down (Hudson, 2013). Her practice is a good example of one that goes beyond definitions and conjures a third way, and where the activist practices that psychogeography belongs to can also encompass activist art that transforms place, rather than simply perceiving it.

In Boston, the Minimum Wage Art group makes totem animals from scrap metal and junk, and places them wherever in public they need to go.[5] An Elephant of teenage proportions appeared in the community gardens at Davis Square, and was eventually removed by irritated gardeners from the new condominiums there. A Rhinoceros was driven to the lawn of Boston City Hall, and crazed totem poles were placed at various points across the city they needed to alter. Meanwhile in Stockholm, there was the appearance of the Stureplan Hyenas. Made by an artist going by the moniker 'Olabo', these (life-sized) concrete hyenas, with real hair tails and red glass eyes, were installed with exactingly professional craftsmanship into the sidewalk (Olabo, 2012). The hyenas possessed a sonic aspect as well, since something embedded in their hairy tails or concrete bodies made the fleets of pure-bred dogs that were walked through the park every day go mad with furious indignation at the beasts. They were accompanied by an official looking plaque, except the city's patron saint was adorned with an eye-patch. The entire operation was made to poke at the 'party-arrangers' in the posh neighbourhood, soaked in status-seekers, where royalty goes to party. Hyenas eat the offal of better animals, and the party-arrangers' activities, combined with the massive, multi-national gentrification of the area, had all but erased any other stories there until the Hyenas came calling.

Walking the full monty

In the soundwalking practices that grew from Schafer's and Oliveros' original writings, there is the commonality that one is a careful and respectful listener. The activism of these pursuits is one of subtle perceptual shifting. It is an activism that is often carried out as an implied, rather than overt, activity. One is an internal agitator, transforming space through internal acts of minute creative rebellion. But the city I walk and listen to is neither an object of ecological revulsion nor a template for the politicized breakage of banality.

Between and outside these poles unfold innumerable inter-worlds, offering a deeper sense of place from sounds that might be shunted or refuted in the origins of these practices. Between 2008 and 2020, the late Ian Rawes created the *London Sound Survey* website, a staggering collection of field recordings from the last century of that city. Neither the term soundwalking nor the term psychogeography appear anywhere on the site; Rawes began making recordings of London in 2008, at the same time as he was working as a storeman at the National Sound Archive. Working with archival recordings and other recordists, he built a vast repository illustrating the history of and present London in maps of sound (*London Sound Survey*; Rawes, 2008–2020). This is another example of a third way, in which elements of both soundwalking (through Rawes' recording practice) and psychogeography (through his re-mapping of the city in a *dérive* of its time-decoupled sounds) are present, but distant enough to offer traversals beyond their particular agendas.[6] The *London Sound Survey* implies that such pursuits could be taken up in other cities, by virtue of its wholly independent production, and its sheer breadth. The site includes sounds that were banned as being a nuisance, like singing or organ grinders, machines, industries and even a 'bloody squawking parakeet' (*London Sound Survey*; Rawes, 2008–2020), as part of its 'orchestration of the soundscape' (Schafer, 1977). Certainly, Rawes needed to listen hard to his surroundings. But is his journey through the city any less transformative for not being attached to the agendas of various walking or listening movements? Rawes' sonic map is something far beyond the all-too-usual button collection of ear-witnessing; his version of London is a third way between and beyond these poles, and its current sonic landscape is richer for this triptych of past and near-present sonic days.

In the soundwalks of both Schafer and Oliveros, humans are eternal visitors. But in the human-built city, one is always a participant, whether permanent or fleeting, living there or trapped there. Wherever one stands in the colliding masses of unjust hierarchies that criss-cross the cityscape, everyone still emerges in the same space, inside the beautiful noise, as co-creators. In the 1940s, Tony Schwartz started recording his own neighbourhood in the 19 zip code of New York (Schwartz, 1954). Pre-dating both the situationists and the acoustic ecologists, his walk in the city was one of curiosity, a search for community and a desire to share the life of his own neighbourhood in sound. He recorded street callers, construction workers, children, machinery and other sounds that tell the story of a myriad of beautiful noises emerging from obstacles, including his own voice in conversations with his subjects. We know about Schwartz's recordings because LPs were issued of them by Smithsonian Folkways, capturing a New York long since flown into the past, in daily moments of the streets that have been erased by the steady march of exploitation. The *London Sound Survey* also saw some of its recordings selected and released on LP in 2013, albeit by a small, now defunct label in London (London Sound Survey, 2013). In this way, rather than having their soundwalking defined by composers, these recordists gave new streets back to music.

Back in Boston, radio artist and performer Jamie McLaughlin recorded nearly every political action, concert or city council meeting she attended throughout the early 2000s, as well as her subway busking performances with an almost magically altered drum machine, moments in the streets and impromptu interviews with visiting state dignitaries she noted were coming to town.[7] The day of the 2000 Boston BioDev protest, she had other obligations, so she asked me to take the minidisc recorder and get the speeches and protests in the streets. This was the very first time I had used my recorder for anything other than very occasional concerts (something I felt queasy about, since I believed improvised music should disappear at its conclusion), and even more occasional recordings of my violin transposition wanderings. In 2004, I bought my first cellular device, a 'dumb phone' with a simple, low-quality recorder. The first recording I ever took of the city at large used that phone. It was of a woman on public transport, who clearly wanted the entire bus to hear her conversation. She drew out every word in a rasping, almost mournfully brash voice, turning the minutia of picking up her niece from school and going to the food stamp office into a frog-throated epic. Another day, I found broken toys in the Goodwill Store, whose electronic sounds warbled and failed in ways sure to alarm any child who might try to play with them. Goodwill stores, poky city buses and inadvertent raspy songs about the food stamp office are an important part of the cities I live in, just as much as the blue jays, whose picturesque bird calls are a warning of intruders, the oak trees along the Charles River, whose traffic-induced illness gives their sickly bark a moonlit, elfin sheen, or the bucolic gardens at Mount Auburn Cemetery, which was so lovely it became the inspiration for many urban parks, with the idea that such gardens might also be made for the living.

McLaughlin played her recordings each week on a pirate radio broadcast which I sometimes joined for, the radio waves accompanying her walks through the city's protests, concerts, meetings, street occurrences and political figures she waltzed up to with the minidisc player and questioned. We broadcast every Saturday, before taking pots of soup and rice down to the Boston Common for Food not Bombs. That weekly walk through the city, including the subway ride, with the rattling shopping cart we'd 'borrowed', complete with the sonic barrage of angry transit authority workers scolding us for taking a shopping cart onboard, and the din of cars, drunk people, tourists, children, church bells and subway machinery around the Boston Common – these are also a part of my sonic orchestration of that city, and of that city's sonic orchestration of my memory. And what reflective listening practice could equal that of listening to the stories told by those who ate our soup, many of whom had seen the same wars we protested, and needed to be listened to with suspended ideology and judgment? Meanwhile, as I walked with my violin on other days through Boston, I soon learned that the traffic boxes, bridges, airplanes, cars, trains and ventilation systems I played with were utterly defenceless. Even in all their clattering, roaring glory, they can never listen in return. Who will take care of them?

In the Square

At Fridhemsplan subway station in Stockholm, I stop at the top of yet another escalator. All the subway escalators have their own knocking boogie. But here, an unoiled stair shrieks out a call to the spirits. I record it, and put it in an installation for Cyklopen, an anarchist-leaning cultural centre. Their wooden house was burned down by right wing extremists, but then rebuilt in a miraculous, colourful design of non-flammable materials. The exuberance of the free jazz escalator sings for perseverance, in the march to and from work, in the Friday night storm of partiers, in the late-night trudge of the lonely.

In 2002, I found Max Neuhaus' *Times Square* by accident. I listened without understanding that I was listening to a work of sound art. I stood on the precarious metal grating and listened to what I believed was a distant machine for a long time, only learning that my sonic companion of that hour was an intentional one nine years later, when I began studies at the Royal College of Music in Stockholm. I had learned to listen out for such things when I found a similarly beautiful ventilation shaft near Harvard University one autumn. It sounded like some choir of river fairies were singing under the stately red brick, covered in shimmering gold and red leaves. You had to stand close to listen; in the sea of sounds cities are often filled with, this one was a choice, and a secret. I felt a mythology of messages in the convergence of those two moments, hearkening to these underground choirs. I was almost disappointed when I learned the one in New York was an intentional work of art.

These singing ventilation grates were also the cure for a different singing that had always filled me with foreboding. I was raised in an aluminium-sided tract-house, in a neighbourhood with only five house layouts, bordered on one side by I-94. Late nights the distant, nightmarish sound of semi-trucks wailing down the highway was a mean song of cocaine-addled drivers working too many hours, flying through the ever-repeated desolation of midwestern suburbia, invading my bedroom as they screamed towards Detroit. Interstate highways, for all their hitchhiking mythologies, are one of the least human-traversable environments. Walking on them is taking your life into your own hands; it is poignant to reflect on the fact that the Peace Pilgrim, who spent 28 years traversing the United States on foot in the name of peace, was killed walking on one (Peace Pilgrim, 1991). In larger cities they run above or below the real life of the streets, and traffic is so dense that the spectre of these massive vehicles, sometimes hitched even more terrifyingly together in twos and threes, is muted by traffic jams. Perhaps some find the rattle of subways jarring, or even consider the sound an inescapable nuisance. Marcia Epstein calls traffic noise 'the other air pollution' citing WHO studies about its ill effects on urban populations (Epstein, 2020, p. 14). But for me, the choirs of the public transit ventilation were a panacea to the desolate midwestern highway I'd fled the world of, singing to me that I was safe on my feet, that the subways were running, that the city was, at least somehow, still there.

Disappearing orchestras

I lived in West Philadelphia from 2006–08. One of the stranger sounds to blanket the late afternoons was the warning signals near a vast field of refineries and the Dupont research facility. They were slow and distant, like a dream of air-raid sirens, entwining with the summer breeze, slow-squeaking trollies and slower traffic on the Saturday afternoons they were tested. The neighbourhood was also visited daily by the Mister Softee ice cream truck, with its particular music box melody. Sometimes it could be heard late at night, prompting one to wonder what else Mister Softee might have for sale. Some evenings, people from the neighbourhood gathered around a car or small truck with prepared beats, and did impromptu hip-hop. On Sunday afternoons in spring, street trombone orchestras appeared, with members ranging from young boys to elderly men. They played music not so unlike that of a New Orleans street band, with passers-by spontaneously dancing in the street. Peppered throughout this soundscape was the occasional, ominous sound of gunfire. There was the song of the cardinal, the street caller selling home-made sweets, or hawking a political revolution, the trollies rumbling down Baltimore Ave, the cars and their blaring music of various ilks, the children who still played in the streets of that neighbourhood, the bells of the churches, the call to prayer of the mosques, on streets festooned with marvellous tiny flower gardens, wafting ganja smoke or anarchist collective houses, their air thick with garlic and soup, all about to fall away to gentrification in a few short years hence. All these things and more were the orchestration of the hours-long walks I took through the city, drinking in the urban, industrial and natural in their tangled up environs, the violent and the loving, the rough, the gentle and the raw: the human, which is what anyone making a soundwalk must also be.

In 1991, I went every day to the Arboretum in my home town. I walked and listened to the river and the woods, in concert with the massive air-duct systems of the University Hospital, the distant train whistles, the traffic on the roads to one side of the river and the lawnmowers in the yards of the well-heeled houses on the other. I went in snow and in rain, day and night, walking. There I met a Catholic mystic. He rode about town on a moped, where he had carefully painted the word 'pray' in loopy, 1970s letters. When I met him on the street, he would pull out handfuls of newsletters about sightings of the Virgin Mary. But in the Arboretum, he told stories about otters, turtles, squirrels and robins. He railed against the increasingly officious fixing up of the park, which he held destroyed the natural environment and upset the animals. Of course, the Arboretum was as artificial as the rest of the small city. It was a garden, planted carefully over a century. The Huron River was long since hopelessly polluted, its fish an over-populated product of the trouble, met by cardinal, starling and jay; the raccoons and squirrels; the lonely walkers, students and townies.

One day I was sitting by the river and the mystic appeared. He pointed out a line of stones that went all across the water, and said he had built it. He'd taken up a stone and seen a rose quartz cross in it, and took this as a sign that he should transform that riverbank into a place for meditation and prayer. So he laid stones across a naturally occurring jetty, to form a sacred heart. He walked into the water every so often, listening to each stone, tuning each in its place to form what he named (as I recall): *A Whitewater Symphony to the Glory of the Virgin Mary and her Son our Saviour, Jesus Christ.* The same year, I went there with a friend who was leaving town for boarding school, and we placed our hands in the water at the full moon to ask for guidance – something she had read in one of her books on mythology, symbols and Jungian archetypes. The sound of the mystic's stones sang both these conflicting spirit worlds to me, melting together with the gentrification of the town, and the inexorable fact of leaving one day. I walked there often, listening to that human-built water, accompanied by the mourning dove in the fog, trains over the bridge, dogs in the sun, robins in spring, ice flats in winter.

A way

FIGURE 9.3 Solna Centrum Fiddler Mural. Photo: Katt Hernandez, 2018. All rights reserved

In walking again through these stories, I aim to show how the way of walking and listening I follow needs not to be named beyond my own individual experience, which I have offered here as one possible way of many. There is no category, like that of 'soundwalk' or 'sonic meditations', that really addresses the openness to the entirety of the sonic environment I am making the attempt to listen to, interact with, affect perception of and consciously draw materials from (Figure 9.3). Neither is the span of practices in psychogeography wide enough to take in what I do when I walk and listen. Urban and Outer-Urban places are too imperfect to fit agendas, too broad not to flummox aesthetic concerns, too varied to be boxed in. There is no practice big enough to encompass them. There is no ideology or composition wide enough to contain them. These named practices can act only as tools, which can be inspiring, but which ultimately must be jerry-rigged to be useful. It is in listening to the conflicting realities of the whole span of places, and ultimately breaking free of the origins of the practice of the soundwalk, that its real gifts can be attained.

Notes

1 I performed there with video artist Walter Wright on whom more information is available here: http://experimentaltvcenter.org/walter-wright-and-his-amazing-video-machine; bassist John Voigt; movement artist Joe Burgio; and guitarist Marc Bisson.
2 I wrote about this practice extensively in an article for *VIS* journal in 2020 (Hernandez, 2020).
3 Some examples discussed in this chapter include Solnit, 2001, 2006; Hudson, 2013, 2016; Self and Steadman, 2007; Macfarlane, 2013.

4 More about Geraldine Hudson can be found at the artist's website: https://geraldinehudson.org
5 More about Minimum Wage Art can be found here: https://www.facebook.com/search/top?q=minimum%20wage%20art
6 Jacek Smolicki also wrote about this kind of pursuit in his PhD thesis. His term for this is 'para-archiving'. More can be found in Smolicki, 2017.
7 More about McLaughlin's work can be found on her website, at https://jammx1.com

References

Boston BioDev. (2000) [Protest] [24–26 March 2000, Copely Square, Boston]. More information available at: https://social-ecology.org/wp/category/biodev/2000boston/

Chtchnglov, I. (1953) *Formulary for a New Urbanism*. Library of Nothingness. Available at: http://library.nothingness.org/articles/SI/en/display/1

Debord, G. (1955) *Introduction to a Critique of Urban Geography*. University of Vermont. Available at: https://www.cddc.vt.edu/sionline/presitu/geography.html

Elkin, L. (2016) *Flaneuse: Women Walk the City in Pairs, New York, Tokyo, Venice and London*. New York: Farrar, Straus and Giroux.

Ende, M. (1979) *The Neverending Story*. Translated by the German by Ralph Manheim. London: Dutton.

Epstein, M. (2020) *Sound and Noise: A Listener's Guide to Everyday Life*. Montreal and Kingston: McGill-Queens University Press.

Gandy, M. and Nilsen, B. (eds.) *The Acoustic City*. London: Jovitis.

Hernandez, K. (2020) Aural Transposition, Psychogeography and the Ephemeral World. *VIS Nordic Journal for Artistic Research* 4 (14 October 2020). Available at: https://www.en.visjournal.nu/aural-transposition/

Hudson, G. (2016) *The Peripheral* [Exhibition]. [Candyland, Stockholm, 3 April 2016]. More information available at: https://candyland.se/?p=778

Hudson, G. (2013) *Human Nature, Pt. 2* [Performance]. [Högdals toppen, Stockholm, Autumn Equinox, 2013]. More information available at: https://geraldinehudson.org/human-nature-pt-2-ritual

London Sound Survey (2013) *These Are the Good Times*. [LP] Vitelli.

Macfarlane, R. (2013) *The Old Ways*. New York: Penguin Random House.

Malia, M. (2003) Transcriptions [Sound Art Installation] [Boom Box: The Art of Sound, Boston Center for the Arts, Boston]. More information available at: http://web.archive.org/web/20021210100653/http://www.bcaonline.org/exhibitions.htm

Neuhaus, M. (1977) *Times Square*. [Sound installation]. More information is available at: https://www.max-neuhaus.estate/max-neuhaus/soundworks/permanent/

Novak, D. (2015) A Beautiful Noise Emerging from an Obstacle. In M. Gandy and B. Nilsen (eds.), *The Acoustic City*. London: Jovitis.

Olabo. (2012) *Hyenas at Stureplan*. [Guerrilla Sculpture Installation]. [Engelbrechts plan, Stockholm, Summer 2012]. More information available at: http://www.streetartutopia.com/?p=3506

Oliveros, P. (1971) *Sonic Meditations*. New York: Smith Publications.

Peace Pilgrim. (1991) *Peace Pilgrim: Her Life and Work in her Own Words*. Santa Fe, NM: Ocean Tree Books.

Rawes, I. (2008–2020) *London Sound Survey*. Available at: https://www.soundsurvey.org.uk

Schafer, R.M. (1977) *The Soundscape: Our Environment and the Tuning of the World*. Rochester, VT: Destiny Books.

Schwartz, T. (1954) *New York 19*. [LP] Washington, DC: Smithsonian Folkways.

Self, W. and Steadman, R. (2007) *Psychogeography*. New York: Bloomsbury.
Smolicki, J. (2017) *Para-Archives: Rethinking Personal Archiving Practices in the Times of Capture Culture*. Malmö: Malmö University.
Solnit, R. (2001) *Wanderlust: A History of Walking*. New York: Penguin.
Solnit, R. (2006) *A Field Guide to Getting Lost*. New York: Penguin.
Takemitsu, T. (1995) *Confronting Silence*. Lanham, MD: Scarecrow Press.

10

'OUR VOICES REACHED THE SKY': SONIC MEMORIES OF THE ARMENIAN GENOCIDE

Gascia Ouzounian

In a letter dated 16 August 1915 and marked as having been 'conveyed beyond the Ottoman frontier by an Armenian refugee from Cilicia in the sole of her shoe' (Bryce, 1916, p. 20), we encounter an agonizing plea from a genocide survivor: 'In haste and in secret I seize this opportunity of bringing to your ears the cry of agony which goes out from the survivors of the terrible crisis through which we are passing at this moment,' she writes. 'They are exterminating our nation, mowing it down. Perhaps this will be the last cry from Armenia that you will hear; we have no longer any fear of death, we see it close at hand, this death of the whole people' (ibid., p. 20).

These tragic words—smuggled out of the Ottoman Empire in a shoe, presented to the British House of Parliament in 1916, and subsequently published as part of a 700-page volume of state papers documenting what would later become known as the Armenian Genocide (Bryce, 1916)—are unambiguous in relaying the scale of the crisis underway: an entire nation is being 'mowed down,' 'exterminated.' The unnamed writer evokes the horror of the moment by rendering it in a sonic register: she seeks to bring Armenia's 'cry of agony' to the ears of the reader. 'We are waifs who cry for the lives of our brothers,' she continues. 'These lines cannot describe our misery; it would need volumes of reports to do justice to that' (ibid., p. 20).

Still, in two short pages, the refugee from Cilicia describes the atrocities already underway in the nine or so months since the Genocide began: deportations and death marches wherein 'tens of thousands' of 'deported widows and children' have been on the road for 'from three to five months,' 'marched along naked and starving'; the wholescale murder of men and boys, who have been 'massacred or drowned'; the abduction and rape of young women and girls, 'three-quarters' of whom have been 'carried off, with the result that one does not see a single pretty face among the survivors'; and the 'Black famine' that plagues survivors, '60 per

DOI: 10.4324/9781003193135-11

cent' of whom 'are sick.' 'The whole of Armenia is being cleared out,' she closes her letter by stating. 'I sign this letter with my blood!' (ibid., p. 20).

In dialogue with this desperate attempt to render audible the pain of a nation, I wonder what it would mean for the cries of a nearly exterminated people to be heard over a century later: for the nearly extinguished, unheard, and concealed voices of genocide victims and survivors, as well as the sounds and sonic violence committed by perpetrators, to be excavated and listened to. Where do these sounds and voices reside? In which bodies, archives, territories, and memories do they continue to sound, and how can they be received, retrieved, heard, and shared? What could such an historical listening reveal about experiences of genocide and the role of sound and listening in its production of trauma?

To better understand the Armenian Genocide and how its traumas were experienced by victims and survivors, I seek here to listen to survivors' sonic memories of the Genocide: what they heard, and how they felt, interpreted, and remembered what they heard. I use survivors' 'earwitness' testimonies—testimonies of auditory and sonic experiences of the Genocide—as a basis for this listening (Schafer, 1994[1977]; Birdsall, 2009). Visual evidence predominates in studies of the Armenian Genocide as well as in the collective imagination of the Genocide. The public spaces of the Armenian Genocide Museum-Institute in Yerevan, the most important public site of genocide remembrance, for example, are almost exclusively devoted to visual displays and to interpreting photographic evidence. By comparison, sonic memories of the Genocide, which carry enormous emotional and affective weight, as well as historical and cultural significance, have generally been neglected as a site of research.

In engaging with sonic memory as a form of sense-making and a site of knowledge, I am particularly indebted to the work of Carolyn Birdsall, who examines the role of earwitnessing and the role of sound 'within personal and social contexts of remembering' (Birdsall, 2009, p. 169). Birdsall moves beyond the usual sonic archives of sound recordings, music, song lyrics, and sound technologies in seeking to recover and examine the audible past. Interviewing elderly residents of Düsseldorf, Germany about their auditory experiences of bombing during the Nazi era, Birdsall engages oral testimony as a site of earwitnessing, investigating 'how remembering and witness testimony are informed by auditory experience' (ibid., p. 170). Whereas, she writes, 'the common conception of the *eyewitness* upholds the notion of an observer who experiences and remembers in visual and semantic terms,' Birdsall investigates modes of sonic remembrance that move beyond 'the visualist associations with witnessing' (ibid., p. 169).

Engaging with earwitness testimony can profoundly reshape our understanding of the Armenian Genocide and its effects, since experiences of the Genocide extend far beyond what was visible, what was seen, and what victims and survivors were able *to* see. Indeed, as becomes clear from earwitnesses' testimonies, many survivors primarily or only had access to auditory experiences of certain traumatic events. Since those who were targeted were often hiding from the perpetrators of genocide, in many cases they had no choice but to interpret the events unfolding around them primarily through listening. And, even in those moments when they *were* able to see (or engage in multiple modes of sensing and sense-making), a sonic memory sometimes emerges as the primary memory of an event. Further, in some cases, a memory of the Genocide has emerged *as sound*: for example, as an unwanted sound that continues to 'resound in the ear' of the survivor. In other cases, a collective memory has taken shape *as song*: for example, in genocide songs that survivors sang to collectivize the pain of the Genocide and to transmit knowledge of an event that was forbidden to speak about publicly and therefore to 'know.'

Sonic memory is a site of historical, cultural, and affective knowledge; and sonic memories can be formed through individual as well as collective and shared processes of remembering. Listening to sonic memories can thus deepen our understanding of the Armenian Genocide both in terms of its historical dimensions—in better understanding what occurred—as well as in connection to its social, psychological, and emotional realms. Attending to sonic memories can also be a form of what I call 'counterlistening' (Ouzounian, forthcoming): listening *against* official narratives of the Genocide, and in this case, against the narrative of genocide denial that continues to be maintained by the Turkish state.

Introduction to the Armenian Genocide

Despite over a hundred years of genocide denial by the Turkish state—an official state policy I return to later—the contours of the Armenian Genocide have been well established and are widely agreed. Starting in 1915 and continuing with full force until 1917, with smaller massacres occurring until 1923, approximately 1.5 million ethnic Armenians living in the Ottoman Empire (present-day Turkey) were killed outright, or else died of hunger, thirst, exhaustion, disease, or suicide, as the result of a state-led campaign to rid the Ottoman Empire of its Christian subjects—principally, Armenians, Greeks, and Assyrians—in a bid to consolidate state power and render Turkey 'for the Turks' (Bryce, 1916; Morgenthau, 1918).

Hundreds of thousands of Armenians were killed outright by Turkish gendarmes and Kurdish peasants who received orders to rid the empire of 'infidels' (*gavûrs*, in Turkish). According to witness testimonies, victims were killed with rifles, knives, swords, clubs, axes, daggers, shearing scissors, and truncheons; they were hanged, drowned, beheaded, and dismembered; thrown into pits, wells, rivers, and valleys; buried alive, boiled alive, skinned alive, and burned alive.[1] A common method of mass killing was to gather the Armenian population of a town into barns or stables, soak the building in kerosene, and set it on fire. Survivors recall hearing the anguished cries of those who were burning alive, as well as the celebratory songs of the perpetrators. The survivor Shogher Tonoyan (b. 1901, Mush), who lost 60 members of her family in one such fire, recalls a harrowing scene whereby she and her cousin escaped a burning stable when its roof collapsed by 'treading on burnt logs and corpses' and coming out through a breach. 'There we saw the Turkish soldiers dancing in a circle, swinging and striking their sabres and singing merrily "*Yürü, yavrum, yürü!*" (Dance, my child, dance!)' she recalls. 'Up to this day that song resounds in my ears' (Svazlian, 2011, p. 98).

In seeking to empty Armenian towns of men and thereby weaken resistance, Armenian men aged 25 to 60 were called to military service but were disarmed, shot, or forced to do hard labour (Bryce, 1916, p. 9; Morgenthau, 1918, pp. 302–303). At the same time, prominent Armenian men—industry leaders, religious leaders, and community leaders—were imprisoned under the false premise of anti-Turkish activity. In prison, many were tortured and executed. The use of *bastinados*—the caning of the soles of the feet with a thin rod until the feet 'swell and burst,' often requiring amputation—was reported in numerous districts (Morgenthau, 1918, p. 306). The genocide survivor Mikayel Mkrtich Chilingarian (b. 1904, Harput) says of his father, who was imprisoned for around one month in May 1915, and who was permitted to come home for one night to change his clothes, that 'He was completely changed, he had grown thin and pale; he walked cautiously and slowly, as if he was not walking on his own feet' (Svazlian, 2011, p. 269). Chilingarian recalls, 'My father told us in a trembling voice: "All this is the result of the terrible daily beating; look carefully and never forget it." Up till now, when I recall my father, I remember with horror that scene and his last words' (ibid., p. 269).

Survivors report that Turkish gendarmes cut open the bellies of pregnant women in a 'game' to determine the sex of the baby (ibid., pp. 176, 216, 241, 292, 313, 344, 378, 387, 392). Many survivors also witnessed the beheading of children and adults, remarking that gendarmes used the decapitated heads 'as a football' (ibid., p. 353, 360) or threw them into pits (ibid., p. 432). The survivor Gyurdji Harutyun Keshishian (b. 1900, Zeytun) testifies:

> They made us walk. On the way, they killed women, too, for our turn had come. And do you know what they did? There was a big pit in the desert; they were cutting the heads of women and children and throwing them into the pit. The poor children were dying like lambs .... When the gendarmes finished their work, they poured petrol in the pit to burn the corpses. The dead did not feel, but the voices of those alive would tear your heart into pieces.
>
> *(Svazlian, 2011, p. 432)*

Those who were not killed outright were forced to march to the deserts of Der Zor (present-day Syria), where the average temperature is 40° Celsius in summer and 2° Celsius in winter, thus ensuring death by exposure (Dolbee, 2020). During these death marches, Turkish gendarmes rode on horseback, keeping Armenian deportees, who mainly travelled barefoot, in line with rifles and whips. Survivors testify that they were 'driven like sheep' (Svazlian, 2011, pp. 244, 259, 274, 304, 313, 352, 389, 432, 445, 448). Hundreds of thousands of Armenians died of starvation, thirst, disease, and execution on these death marches. The survivor Ashot Ohanian (b. 1905, Bursa) says, 'We were walking on foot for days and weeks. Our feet were bleeding. The policemen were beating us with whips. Many could not endure the sufferings and died on the road' (ibid., p. 399).

Those who survived the death marches were typically relocated to refugee camps and orphanages where many, too, succumbed to disease. Cholera and typhus were pervasive, as were dysentery, trachoma, pneumonia, and diseases borne of malnutrition. One orphanage in Alexandropol (present-day Gyumri), managed by the charity Near East Relief, housed 25,000 children. A survivor from Moosh recalls the moment her group arrived at Alexandropol, travelling on foot. They found that 'thousands of refugees had gathered and typhus was devouring innumerable victims' (ibid., p. 96). Her group set off to Tbilisi, Georgia—some two hundred kilometres away—the very next day.

During these tragic years, some Armenians committed suicide to escape murder, rape, or forced Islamization. Some survived by marrying Muslims or by being 'adopted' by Muslim families. These survivors were forced to renounce their Armenian-ness and their Christian religion, and to discontinue speaking the Armenian language. Many Armenian women and girls were sold into sexual slavery, including by public auction, and forced to join harems (Bryce, 1916, p. 7). Photographs of their tattooed faces line the walls of the Armenian Genocide Museum in Yerevan today.

Despite the measures taken by the Ottoman government to suppress communication with the outside world (Suny, Göcek, and Naimark, 2011), news of the events circulated globally at the time they occurred, including through detailed reports and testimonies by diplomats, journalists, doctors, humanitarians, missionaries, travellers, survivors, and witnesses; through photographs that were smuggled out of the empire; as well as through articles in numerous international journals and newspapers.

While most historians agree that the deaths of Armenians and other Christian subjects in the Ottoman Empire were the result of a state-led campaign of mass extermination, the Turkish state maintains that these deaths were the result of a military strategy of 'relocation' and, in the Armenian case, a justifiable response to Armenian separatism and nationalism. In present-day Turkey, it is considered a crime of 'insulting the Turkish nation' to recognize the Armenian Genocide as such. This crime carries a sentence of six months to two years in prison according to Article 301 of the Turkish Penal Code.

The denial of genocide by the Turkish state is not only notable in how long it has persisted in the face of incontrovertible evidence and international pressure, but also in that the principal orchestrators of the Genocide, known as the Three Pashas—members of the Young Turk party and leaders of its far-right Committee of Union and Progress—were sentenced to death by Ottoman Military Tribunal in 1919 for bringing the empire into war and 'organizing massacres against Armenians and Greeks' ('Turkey Condemns Its War Leaders,' 1919; see also Suny, 2015). Given this early recognition of war crimes by the Turkish state itself, the denialism that followed can be thought of as part of a larger project of nation-building, most notably marked by the replacement of the Ottoman Empire (a sultanate) by the (constitutional) Republic of Turkey, established in 1920–23 under the leadership of Mustafa Kemal Atatürk, who served as its first president until his death in 1938.

Soundwalking through history

In seeking to trace what survivors heard during the Armenian Genocide, and how sonic violence formed part of its injuries (and violent tactics), I have consulted contemporaneous news articles and diplomatic and humanitarian reports, as well as secondary genocide literature. My main source here, however, is the essential volume of survivors' testimonies collected over a 50-year period by the Armenian ethnographer Verjiné Svazlian and published in its English translation in 2011 as *The Armenian Genocide: Testimonies of the Eyewitness Survivors* (Svazlian, 2011). Starting in 1955, when, as she stresses, 'it was not possible to speak explicitly about the Armenian Genocide in Soviet Armenia,' Svazlian went from village to village, recording and transcribing verbatim some 700 survivors' 'memoir-testimonies,' including 315 'song-testimonies' or genocide songs: Armenian and Turkish-language songs that memorialized the events of the Genocide, and that some survivors spontaneously sang as they recalled those events.

We can think of Svazlian's act of collecting testimonies—one that she began while still a university student and continued over a period of decades despite the significant personal risk—as an original act of counterlistening, and even as a form of soundwalking. 'In [Soviet] Armenia, under the scorching summer sun and in the icy winter cold,' she recalls, 'I went on foot, from district to district, from village to village, searching and finding eyewitness survivors miraculously rescued from the Armenian Genocide. I approached them tactfully, without diverting their attention with irrelevant questions, and let them freely express their immediate impressions. I wrote down (and also tape-recorded) the bewildering memoires, the impressive stories and the diverse historical songs, which they narrated and sang' (Svazlian, 2011, pp. 7–8).

Svazlian's process of collecting testimonies was therefore rooted in walking and listening. She had no choice but to walk from village to village, since many survivors, dispossessed and living on the margins of society—'plundered, left destitute' (ibid., p. 13)—resided in remote areas that were difficult to reach in any other way. She also walked because of the radical, dissident nature of her work, which would have prevented a more public form of movement and research. Svazlian's process was equally rooted in focused listening: she listened, as she says, without diverting survivors' attention. Indeed, it is only because of Svazlian's original, self-sacrificing act(s) of listening that we can now 'listen' in a collective, shared, or public way. It is notable that Svazlian's sensitive approach to listening enabled survivors to speak *and sing*, given that many had never previously shared their experiences of the Genocide. Her listening enabled survivors' memories to come to the surface, be transmitted through speech and song, and through the recording, transcription, publication, and circulation of these memories, it has enabled us, too, to listen.

Svazlian herself uses a sonic metaphor to describe the force of emotion that arises from concealed and denied memory. The work of collecting survivors' testimonies 'has no end,' she writes, since 'there is not a single Armenian family, who has not suffered human and material losses as a consequence of the Armenian Genocide' (ibid., p. 9). 'For that reason,' she continues, 'the theme of the Genocide raises its voice and roars in the blood of the Armenian people' (ibid., p. 9). In attending to survivors' sonic memories, my aim is to hear this voice-in-the-blood, and for this voice to leave the blood and enter the air.

Earwitnessing genocide

In engaging with a deeply contested history—one that, if addressed publicly in present-day Turkey can cost one's freedom or even one's life (Freely, 2007)—it is important to convey from the outset that I regard survivors' testimonies as sacred utterances whose truthfulness is sacrosanct.[2] I do not reference survivors' testimonies as 'proof' of the Armenian Genocide, even as they serve, in my opinion, as its most important body of evidence. As a descendant of genocide survivors (each of my grandparents or great-grandparents was a survivor), I do not qualify so-called 'debates' on the Armenian Genocide by engaging with them as such. Such debates are only alive because of state denial, and I see them as a continuation of the violence perpetrated by the Turkish state towards Armenians.

Still, I do wish to reflect on the special *quality* of truth that emerges through survivors' testimonies, both individually and as a collection. On the one hand, an evidentiary or historical truth emerges as similar experiences are reported by people from different regions who could not have had any knowledge of one another. But a deeper, cultural truth emerges as well: a truth about a people who shared a common language, customs, and traditions—ways of feeling and ways of remembering. An endangered (nearly exterminated) culture is transmitted through survivors' testimonies, including through such linguistic phenomena as turns of phrase; through *what* is recalled and *how*; and in the expression of certain kinds of pain and sorrow. In the case of survivors of the Armenian Genocide, this includes the spontaneous recall of genocide songs—what might be understood as a form of sonic resistance and survival: a way of memorializing a traumatic past that has been concealed by the state; transforming that pain into song; and using the medium of song to retain collective memories and ensure their circulation and survival.

Regarding the testimonies collected by Svazlian, it is notable that many are remarkably detailed; one survivor, Karapet Tigran Kelekian (b. 1904, Everek), for example, recounts each leg of the journey he was forced to take on foot as a child of 11 years, 'walking daily eight hours' (Svazlian, 2011, p. 373). He remarks that 'the length of our route was 677 km. But it was planned so that we shouldn't use built roads, so the route must have been at least 800 km long' (ibid., p. 373). Many testimonies are also strikingly imagistic, whether they depict visual or sonic scenes. The survivor Mesrop Manvel Mesropian (b. 1900, Yozgat), for example, describes a scene whereby the Biblical Day of Judgment had 'become a reality.' He testifies:

> Abandoned by God and friends, unprotected in the open field, the widespread horror of death waved like a huge flame driven by a terrific blast over the heads of 10,000 innocent, sacrificial Armenians....
>
> The terrified children's screams, the wail of the mothers whose children had been relentlessly grasped from them, the heart-rending cries of the children who were being taken away from their mothers, the moans and groans of those in agony blended into an awful funeral dirge, which joined the curses and clamors of those who were waiting for their turn to be martyred. Thunderbolts were sent to heaven: 'God, smell our innocent blood and take vengeance.'
>
> *(Svazlian, 2011, pp. 348–349)*

Earwitness testimonies such as Mesropian's go far beyond relaying the *events* of the Genocide. They give voice to textures of experience of genocide, conveying the full force of the brutality that occurred and the depth of the suffering that was endured. The suffering that Mesropian witnessed was manifested, among other ways, in *and through* sound. He heard terrified children's screams; the wails of mothers whose children were being taken from them; the moans of the dying; and the curses of those awaiting execution. For him, those sounds 'blended into an awful funeral dirge' as though each of those tormented voices was a distinct part in a chorus of suffering.

Those sounds also caused him to suffer. They transmitted the torment and suffering of others, and, by earwitnessing them, Mesropian himself felt and 'absorbed' that torment. Attending to traumatic sonic memories such as Mesropian's can therefore not only help to establish the facts of the Genocide, but it can also enable a more nuanced appreciation of how victims and survivors sensed and interpreted its manifold overwhelming traumas.

Announcing exile

The first sounds that many survivors recall of the Armenian Genocide are the voices of Turkish town criers who arrived at Armenian districts on horseback, announcing the orders that Armenians should either convert to Islam or be exiled. Nouritsa Kyurkdjian (b. 1903, Aintab) says, 'One day, we heard horse hoofs from the street. The town crier, Moukouch, announced, "Listen, hey, people! In twenty-four hours, you must either change your homes or your religion"' (Svazlian, 2011, p. 453).

The survivor Yeghiazar Karapetian (b. 1886, Sasun) remembers the voices of Turkish town criers as emerging against a backdrop of silence and the foreboding caws of black crows. 'On Monday morning, the 29th of June [1915], the streets were empty, no human being was seen: there was neither whisper nor movement,' he recalls. 'Only the ominous caws of thousands of black crows were heard from the tall poplars. At the moment when the sun had quite risen in the sky, all of a sudden the shrill voices of the Turkish town criers were heard in the Armenian districts' (Svazlian, 2011, p. 83). The criers ordered local Armenians to gather with their carts, mules, property, and family and report to the local government building; those who refused to obey the order would have their property confiscated and would be 'exiled by force' (ibid., p. 83).

Along with the voices of Turkish town criers, survivors recall hearing the terrifying sounds of cannonade and gunfire as the killings got underway. One survivor recalls, 'At the time when thousands of cannons and guns were thundering over the Armenian quarters, pitiless voices of the town criers were heard from the Turkish districts' (ibid., p. 83). Another survivor, who was seven years old at the time of the events, recalls, 'We heard the cannonade of the Turks. We hid behind rocks. I used to cry out of fear .... The cannonade sounded like thunder' (ibid., p. 161).

Several survivors recall hearing 'drum and zurna' (the zurna is a reed instrument) as Turkish gendarmes descended upon Armenian villages and towns. These instruments were sounded for several reasons: to signal Turkish presence and power, representing a form of sonic territorialization; to accompany the celebratory sounds of Turkish victory over the vanquished—and thus a form of sonic triumphalism (as in the song that Shogher Tonoyan heard as her family burned); and to mask the sounds of torture and murder. In *Ambassador Morgenthau's Story* (Morgenthau, 1918), Henry Morgenthau, who served as the American Ambassador to Turkey between 1913 and 1916, details the various methods of torture that Turkish gendarmes used on Armenian prisoners. 'These cruelties,' he writes, 'were usually inflicted in the night time. Turks would be stationed around the prisons, beating drums and blowing whistles, so that the screams of the sufferers would not reach the villagers' (Morgenthau, 1918, p. 306). In the early stages of the Genocide, it was advantageous not to raise alarm and thereby inspire resistance. Therefore, the sounds of drum and zurna were sometimes deployed as a military tactic: to mask the sounds of suffering and hide genocidal intentions.

As the killings and deportation orders spread, survivors recall hearing 'turbulent noises' (Svazlian, 2011, p. 528) and general 'noise and clamor' (ibid., p. 366). In survivors' memories, these noisy soundscapes are often contrasted with the idyllic soundscapes of village and town life. Many survivors remember their villages and hometowns with profound affection and longing, and associate them with the comforting sounds of rivers, birdsong, church bells, and festivities.

The turbulent noises of the starting genocide are also contrasted in survivors' testimonies with the terrified silence of victims. Heghnar Gabriel Ghoukassian (b. 1896, Pasen, Delibaba Village) testifies, 'We were to pass the canyon without any noise or voice. Not a stone would stir under our feet .... Mothers [were ordered] not to take out their nipples from their babies' mouth, even if the child got choked, so that any crying voice would not be heard. We passed by night, we passed without any noise' (ibid., p. 121). Many survivors report that families were forced to hide or abandon their babies for fear that their cries would attract Turkish soldiers: 'They left the babies under the trees, for they were crying. They did not keep any babies, because the enemy would hear their crying' (ibid., p. 89).

'My uncle's voice stopped'

Among earwitness memories of the Armenian Genocide, memories of voices—whether the shrill voices of Turkish gendarmes, or the weak and fading voices of the dying—are recalled in particular emotional detail and depth. Survivors often specify the *quality* of the voices they heard, conveying the emotional landscape that was wrought through the voice. They describe, for example, 'faint voice' (Svazlian, 2011, pp. 236, 238, 339); 'trembling voice' (ibid., p. 269); 'heavy voice' (ibid., p. 238); 'low voice' (ibid., pp. 262, 315, 357, 375); 'shaky voice' (ibid., p. 321); 'quivering voice' (ibid., p. 334).

In Svazlian's volume, there are over 250 references to 'crying,' and many to 'shouting,' 'screaming,' 'wailing,' 'yelling,' 'pleading,' 'moaning,' 'groaning,' and 'lamenting.' The survivor Souren Sargsian (b. 1902, Sepastia/Sivas) describes a scene in which women 'were crying and lamenting over their children, screaming and cursing their fate and their luck in hoarse voices' (ibid., p. 319). He also describes a scene in which 'thousands of voices' came together in a plea for salvation:

> Eight to ten gendarmes armed with bayonets and fifteen to twenty Kurds surrounded us, made us sit in a sandy mound in rows, according to age and height. A woman came to save her little boy. One of the gendarmes cursed and pushed her roughly into a pit. She got up screaming plaintively: 'Help, they're slaughtering the children!' Suddenly voices of thousands of women rose in the darkness to heaven. The chief of the gendarmes ordered to open fire towards the people. They took the rifles and fired towards the people: no more voices were heard, only the moan of the wounded women.
>
> *(Svazlian, 2011, p. 320)*

Several survivors report listening to fading or disappearing voices as a marker of death—as evidence that a loved one was no longer alive, as in the story recounted by Ronia Terzian (b. 1920, Aleppo):

> In the silence of the night we heard the sound of whipping and my uncle's voice calling, 'Help! Help!' .... My mother was crying, was pulling her hair and was rolling on the ground. Then she lost consciousness. The strokes of the whip were heard for a long time. My uncle's voice became weaker and weaker. I was crying for my uncle and my mother. The neighbors sprayed water on my mother's face. She opened her eyes, said 'Aram, my brother,' and closed her eyes again. My uncle's voice stopped, and the whip strokes stopped.
>
> *(Svazlian, 2011, p. 525)*

As these testimonies show, the voice can be a crucial site of listening to genocide. Victims cried and screamed from sheer pain and terror, but also because they wanted others to hear their pain. The voice was often the only tool victims had to express their suffering. Attending to sonic memories of victims' voices can therefore be a way of recovering these concealed voices, and a step towards recognizing their pain.

'We walked among corpses'

In the sound studies literature, walking is typically posited as a methodology that enables a more active, intimate, and embodied exploration of a sonic place. In this literature, soundwalking is justifiably celebrated as a means towards producing a lived, sensorially rich knowledge of a sonic environment: embodying a sonic choreography of an urban environment by 'walking the sonic city,' for example. By contrast, the stories of walking relayed in survivors' testimonies are of walking as a means of mass extermination, torture, and dispossession.[3] As Henry Morgenthau writes, from April 1915 'despairing caravans' set forth from 'thousands of Armenian cities and villages.' At first, 'the individuals bore some resemblance to human beings'; soon, however, 'the dust of the road plastered their faces and clothes, the mud caked their lower members, and the slowly advancing mobs, frequently bent with fatigue and crazed by the brutality of their "protectors," resembled some new and strange animal species.' Morgenthau attests that, from April to October 1915, 'practically all the highways in Asia Minor were crowded with these unearthly band of exiles. They could be seen winding in and out of every valley and climbing up the sides of nearly every mountain—moving on and on, they scarcely knew wither, except that every road led to death' (Morgenthau, 1918, pp. 313–314).

Survivors recount how, after walking barefoot for weeks while deprived of food and water, many deportees simply lay on the ground and perished. Azniv Grigor Siradeghian (b. 1909), a survivor from Sepastia who was exiled to the Surudj desert, and who was only six years old at the time of the Genocide, set off on the road with her mother and four younger siblings. All died on the road, 'of hunger and exhaustion' (Svazlian, 2011, p. 339). 'For we walked and walked up one mountain, then the next, we went up and down seven mountains,' Siradeghian recalls. 'Mother was carrying her baby; she was so exhausted that she could not walk anymore; she fell down .... We walked every day, we walked among corpses' (ibid., p. 339).

Many survivors of the death marches describe the anguish at being forced to leave their dead and dying relatives behind, and at the horror of seeing so many unburied corpses on the road. The survivor Kadjouni Toros Gharagyozian (b. 1905, Shabin Karahisar) testifies:

> On the road, everywhere, we met devoured and rotten corpses of women and children. We saw there a dead mother and her dead baby; another elder child had fallen on his mother's breast; he was still alive and was writhing and crying in a faint voice. We passed by these pitiful scenes with indifference: thinking that one day we would be in the same state .... People died, succumbing to hunger and the tortures of the road.
>
> *(Svazlian, 2011, p. 236)*

The sounds of the death marches were the sounds of subjection and terror. Survivors recall hearing whip strokes, gunshots, and the yelling of Turkish gendarmes; the screams and wails of the deportees; the groans and moans of the dying; and the 'deafening howling' of wild animals such as jackals and wolves that devoured corpses at night (Svazlian, 2011, p. 387). The survivor Galoust Gevorg Soghomonian, who was born in 1905 in Bolou, and whose father was a second cousin of the celebrated Armenian priest and composer Komitas Vardapet, remembers the moment his family was 'forced to join the multitude of Armenian exiles, of which no beginning and no end were visible' (ibid., p. 389). Moving forward under whip strokes, they learned they would be taken to the Der Zor desert. 'Cries and wails were heard from everywhere,' Soghomonian recalls. Those who fell behind were shot or killed with a bayonet, the Turkish gendarmes saying '"*Geber, gâvur!*" ["Die, infidel"]' (ibid., p. 389).

Soghomonian describes a brutal, terrifying soundscape on the road to exile. Each day began with the yells of the Turkish gendarmes, who woke up the exhausted deportees. He recalls:

> I was, at that time, ten years old and I clung, terrified, to my parents. I felt very anxious on hearing the loud shouts of the Turkish gendarmes, the cries and moans of the hardly moving people, the gun-shots resounding from all directions. I wanted to be near those sounds and see with my eyes everything going on around me; I wanted to help those poor people, but the horror I felt as a child made me cling more and more to my parents' hands. I felt that every gun-shot took a human life away. Those who did not obey orders were pierced with a bayonet or had their head cut with a sword.
>
> *(Svazlian, 2011, p. 389)*

Soghomonian's testimony gives voice to the sonic experience of the death marches from the perspective of a young child. He describes a chaotic acoustic scene that made him 'very anxious': he wanted instinctively to see what was happening around him and help those who were suffering, but he was terrified to do so. As a child, the sounds he heard were larger than life: he heard gunshots as 'resounding from all directions' and felt each one as taking away a human life. The sounds were both all-encompassing and piercing; both overwhelming and paralysing. The feeling that a soundscape was both captivating and traumatizing is repeated across survivors' testimonies. *Not* listening to violent sounds was not an option if one wanted to survive; yet hearing violence and suffering was itself a form of injury and trauma.

Indeed, some survivors testify that during the death marches they 'became deaf' to the pain of others as a form of self-preservation. By mentally blocking out traumatic sounds, they were more able to withstand the trauma of the events unfolding around them.[4] As one survivor says, 'The gendarmes … brought the girls in white clothes. In the darkness of the night, they impaled them all with the sharp stakes. Our ears became deaf to their and their mothers' screams, cries and heart-rending clamors' (Svazlian, 2011, p. 321).

The death marches were a form of mass extermination as well as of dispossession. After all, deportees were being marched *away* from their homes, towns and villages, families, livelihoods, and possessions. Even when the assigned destination was reached, there was a strong chance that the deportee would be killed at that spot—burned alive, killed with swords, or shot and thrown into rivers. It was not a euphemism to say that the waters of the Euphrates and Tigris rivers ran red with the blood of Armenians during these years: they were said to be 'the sepulchre of thousands of Armenians' (Bryce, 1916, p. 14). The survivor Kadjouni Toros Gharagyozian (b. 1905, Shabin Karahisar) recalls a scene at one such final crossing, describing the moment when his group arrived at the Euphrates River:

> Everybody was crying, shouting, screaming. Our voices reached the sky. Women were praying to God for salvation. We stood before two kinds of death—either we had to throw ourselves into the river or be driven by force to the cave to be slaughtered by the hands of the executioners. Someone set an example: the mother of a family gathered her children closely about her, raised her hands, crossed herself and shouted praying, 'Oh Lord, you did not help us, now we ourselves are coming to you. We are innocent.' Many followed her example.
>
> *(Svazlian, 2011, p. 236)*

Water features heavily in survivors' testimonies of the death marches, although sonic memories of water are not of pleasant waterscape sounds as is typical in the sound studies literature, in which water sounds are often cited as a marker of a 'healthy' or desirable sonic environment (Schafer, 1994[1977]). Rather, survivors recall being so thirsty that they drank water out of the footprints of camels (Svazlian, 2011, p. 380) and from 'dirty puddles' in which bugs were visible (ibid., p. 254). Some survivors even recall diving beneath corpses in the Euphrates to drink. Aharon Melikset Mankrian (b. 1903, Hadjn) testifies:

> The water of the Euphrates was bloody, it was impossible to drink it; the corpses floated down the current and we dived to the bottom of the river in order to drink clear water …
>
> Those who could no longer walk, sat on the ground or lay down whispering 'water—water' and died. There were dried corpses all around.
>
> *(Svazlian, 2011, p. 428)*

A prominent memory of a 'water sound' during the death marches, tragically, is of the sounds of people begging for water as they perished, and agonizing pleas for salvation at the river's edge.

Locating traumatic sonic memories

As survivors share their sonic memories of the Armenian Genocide, several use the expression that a sound is 'in their ears.' On one hand, this expression can signal the enduring presence of a memory: a sound is felt as *still present*. A sound in one's ear can also be a form of the past 'calling' to the present, seeming to speak directly to the listener. Heghnar Ghoukassian (b. 1896, Pasen), who recalls her hometown of Delibaba Village with deep affection, says, 'A river ran through the village. I have become old now, sometimes my ears whisper, and it seems to me it's our river whispering. Or, from afar, the noise of our millstone reaches my ears' (Svazlian, 2011, p. 121).

For some survivors, the feeling of sound being 'in the ear' can be a source of connection to the past. The only memory that some survivors have of a murdered relative, for example, is of the sound of their voice. Aram Momdjian (b. 1909, Marash) says, 'I was six years old when they came and drove us to exile. I hardly remember my father .... I remember my mother also quite dimly: only her words are in my ears: "They're taking us to Der-Zor to kill us. At least, let our children live"' (ibid., p. 449).

Sonic memories that live 'in the ears' can be a site of both pain and attachment—the connective tissue between a lost generation and a living one. Locating a sound in one's ear can also be a way of giving memory body: keeping the voice of a loved one alive and close by locating it *in* the body.

For some survivors, however, traumatic sonic memories continue to 'resound in their ears' in an unwanted fashion, or what might be thought of as involuntary sonic recall.[5] As Shogher Tonoyan testifies, the song she heard her family perished continues to resound in her ears 'up to this day' (ibid., p. 98). Such sonic memories can be a site of re-traumatization: painful sounds are recalled involuntarily and, being present 'in the ears,' continue to disturb the listener who has no choice but to hear them.[6]

For Tonoyan, the horror of the Genocide was not only encapsulated in the fire that killed 60 members of her family and many of her fellow townspeople, but also in the sonic triumphalism that accompanied it—a triumphalism that was both manifested in and *produced through* song: a 'merry song' that raised the morale of those who were committing acts of genocide—a song that lubricated their violence, and that continues to torment one who heard it. Through her testimony, Tonoyan transmits an impression of those sounds, which have been inscribed 'onto' her ('in my ears'), and which continue to resound 'in' her. The injury that was done to Tonoyan was therefore not only the violent murder of her family, but also the *sonic imprint* of that crime, which remains. The song she continues to hear decades later is an example of what Mhamad Safa has theorized as the 'aftersound' of a violent event: a sonic aftershock that continues to produce trauma in its unfolding and resounding *over time* (Safa, 2022; Ouzounian, 2022).

Counterlistening to genocide

The value of sonic memory to knowledge is not only evidentiary; it also lies in the emotional matter that sonic memories can carry and transmit. If we seek to understand the incomprehensible—an historical event of such tragic proportions that it confounds human comprehension—we must seek to better understand the nature of those emotions and the role of sound and listening in producing and shaping them. This can enable us to better understand not only an 'historical event' (the concept itself is arguably a misnomer since the effects of genocide are not contained within any violent act but unfold across time and space)—but the conditions of the survivors themselves: their pain, terror, and sense of loss; how they feel, remember, and share traumatic experience; and how those traumatic experiences can emerge in connection to sound and listening.

Sound is both a manifestation or expression of trauma—as in the sounds of screaming, crying, lamenting, and moaning—and it can be traumatizing in and of itself. Sounds can transmit trauma between people through shared affective experience: hearing suffering causes suffering. Listening is not a neutral way of engaging with the world; it can be injurious, even when sounds are remembered and recalled versus heard 'in the moment.'

As earwitness testimonies of the Armenian Genocide show, sound can be deployed as a weapon of mass extermination (as in the gunshots that induced terror in Armenian deportees); as a means of territorialization (as in the sounds of drum and zurna that signalled Turkish power); as a military tactic (as in the gendarmes' use of noise to mask the sounds of torture); and as a way of asserting state control (as in the voices of town criers that carried deportation orders, or the yelling of gendarmes that kept deportees in line). Thus, sound played a key role in *producing* the Armenian Genocide and its traumas.

However, we must not only pay attention to how perpetrators of genocide used sound. Even as they were left to die, victims had voice. The sonic memories of survivors transmit those voices, whether the faint voices of the dying, the anguished voices of the crying and lamenting, or the desperate voices that came together in collective pleas for salvation, mercy, and justice. Listening to sonic memories can enable voices that were brutally suppressed—and that the Turkish state continues to conceal and deny—to sound and to be heard. Through listening to the voice-in-the-blood and the 'voices that reached the sky,' we can hope to better understand the incomprehensible, the unconcealable, the undeniable: that which continues to resound.

Dedication

This chapter is dedicated to my beloved grandparents Smpat Ouzounian, Armenouhie Ouzounian (née Kargodorian), Antranig Hamboyan, and Sirvart Hamboyan (née Margarossian). May they have peace, and justice.

Acknowledgements

This research is only possible because of the immense undertaking of Verjiné Svazlian to bring the testimonies of Armenian Genocide survivors to a wider public. All of us who study the Genocide are in her debt. I am grateful to Jacek Smolicki for the opportunity to contribute to this volume, and for his sensitive and thoughtful feedback. This work benefitted from the encouragement and feedback of family, friends, and colleagues including Sirvart Hamboyan, Zareh Ouzounian, Lena Ouzounian, Maral Ouzounian, Karen Ouzounian, Eve Egoyan, Atom Egoyan, Vazken Khatchig Davidian, Carolyn Birdsall, Samantha Dieckmann, Gerard Gormley, Louis Chude-Sokei, Jan St Werner, Çağla Ilk, and Laudan Nooshin.

This chapter was funded by the European Research Council (ERC), under the European Union's Horizon 2020 research and innovation programme, as part of the project 'Sonorous Cities: Towards a Sonic Urbanism' (grant agreement No. 865032).

Notes

1 These methods of killing are all described in Verjiné Svazlian's *The Armenian Genocide: The Testimonies of the Eyewitness Survivors* (Svazlian, 2011).
2 Survivors of trauma are considered unreliable narrators by some psychologists and sociologists in that they 'often tell their stories "in a highly emotional, contradictory, and fragmented manner which undermines their credibility"' (Herman, 1992, p. 1, cited in Greenberg, 1998). I nevertheless regard genocide survivors' testimonies as an attempt, however fragmented or imperfect, to convey truth—and as an undeniable *will to truth*. By 'truth' I am referring not only to historical or scientific fact, but also the truth of experience—the truth of how individuals and communities make sense of their lived experiences. It is particularly in the realm of experience that sonic memory, as a form of sense-making, can be 'telling' as a site of knowledge.
3 As a result of this painful history and the continued denial of the Armenian Genocide by Turkey, acts of walking, marching, and pilgrimage hold special meaning for Armenians. Each year on 24 April, Armenian Genocide Remembrance Day, Armenians around the world take part in mass marches to commemorate the Genocide and protest Turkish denialism. Armenians have also undertaken pilgrimages to the sites of the death marches, both to commemorate the tragedies that took place there, and to collect evidence, such as bones and skulls, which continue to be unearthed in archaeological digs (Semerdjian, 2018).
4 In recovering traumatic sonic memories, it is important to recognize how deeply painful it was for victims and survivors of the Armenian Genocide to hear the suffering of others—to the extent that some 'blocked out' these sounds to survive. In *Regarding the Pain of Others* (2003), Susan Sontag famously questioned the sensationalist and sensationalizing aspects of war photography, as well as the problematic nature of *looking* at photographs of victims of war, asking 'Who are the "WE" at whom such shock-pictures are aimed?' (Sontag, 2003). Sontag's question is an important one to revisit in the context of 'listening to the pain of others,' because 'we' can listen to other people's pain from a distance. The sonic archive affords that distance, just as the medium of photography does.
5 See Birdsall (2009) on 'involuntary remembering' in earwitness testimonies.
6 Birdsall (2009, p. 178) makes the salient point that, since some earwitnesses re-experience a sound in the present moment, it is questionable whether this should be called a sonic *memory*.

References

Birdsall, C. (2009) Earwitnessing: Sound Memories of the Nazi Period. In K. Bijsterveld and J. van Dijck, eds., *Sound Souvenirs*. Amsterdam: Amsterdam University Press, pp. 169–181.

Bryce, J. Viscount. (1916) *The Treatment of the Armenians in the Ottoman Empire, 1915–1916: Documents Presented to Viscount Grey of Falloden by Viscount Bryce*. London: Hodder and Stoughton.

Dolbee, S. (2020) The Desert at the End of Empire: An Environmental History of the Armenian Genocide. *Past & Present*, 247(1), 197–233. doi:10.1093/pastj/gtz055

Freely, M. (2007) Why They Killed Hrant Dink. *Index of Censorship*, 2, 15–28.

Greenberg, J. (1998) The Echo of Trauma and the Trauma of Echo. *American Imago*, 55(3), 319–347.

Herman, J.L. (1992) *Trauma and Recovery*. New York: Harper Collins.

Morgenthau, Henri. (1918) *Ambassador Morgenthau's Story*. Garden City, NY: Doubleday, Page & Company.

Ouzounian, G. (forthcoming) Counterlistening. *English Studies in Canada*, 47(1–2). Forum on Disciplinary Listening.

Ouzounian, G. (2022) 'Aftersound: A Conversation with Mhamad Safa,' The Showroom London. https://www.theshowroom.org/library/countersonics-radical-sonic-imaginaries (accessed: 13 August 2022).

Safa, M. (2022) Reverberations and Post-war Trauma: The Sustained Aftermath of Aerial Strikes on Lebanon in 2006. *Sound Studies*, 8(1), 73–99. https://doi.org/10.1080/20551940.2021.2021367

Schafer, R.M. (1994[1977]) *The Soundscape: Our Sonic Environment and the Tuning of the World*. Rochester, VT: Destiny Books.

Semerdjian, E. (2018) Bone Memory: The Necrogeography of the Armenian Genocide in Dayr al-Zur, Syria. *Human Remains and Violence*, 4(1), 56–75.

Sontag, S. (2003) *Regarding the Pain of Others*. New York: Farrar, Straus and Giroux.

Suny, R.G. (2015) *'They Can Live in the Desert but Nowhere Else': A History of the Armenian Genocide*. Princeton: Princeton University Press.

Suny, R.G., Müge Göcek, F., and Naimark, N.M. (eds.). (2011) *A Question of Genocide: Armenians and Turks at the End of the Ottoman Empire*. Oxford: Oxford University Press.

Svazlian, V. (2011) *The Armenian Genocide: Testimonies of the Eyewitness Survivors*. Yerevan: Gitoutyoun Publishing House of the National Academy of Sciences of the Republic of Armenia. Translated from the Armenian by Tigran Tsulikian and Anahit Poghikian-Darbinian.

'Turkey Condemns Its War Leaders.' (1919) *New York Times*, 13 July, p. 1.

11

COMPOSING, RECOMPOSING, AND DECOMPOSING WITH SOUNDSCAPES

Jacek Smolicki

Introduction[1]

Intertidal zones are areas where sea meets land. They remain above water level at low tide and underwater during high tide. They are characterized by a multitude of organisms evolutionarily adapted to these cyclically varying conditions. Upon one of many visits to such intertidal zones near a place known as Stanley Park in Vancouver, while carefully making my way through muddy rocks at low tide, I slipped.[2] I fell down, hurting my back, head, and right arm. Blood instantaneously rushed from several deep wounds. Dipped in a mixture of mud, sea salt, and pieces of sea grass, my arm hurt badly. The pulsing pain kept increasing, instantaneously cutting me off from the surrounding environment and its soundscape I thought I was almost perfectly in tune with. Instead, this sudden sense of discomfort and pain sent me inward, into my body, drastically reminding me of its limitations. The tide was gradually taking back the patch of shore I fell on. I slowly put myself together, rinsed the wounds, collected my soaked recording equipment and began walking, towards the sea wall. This time, extremely carefully.

Some sites do not want our presence. With some places, human bodies are essentially incompatible, regardless of how strong our efforts to augment them. While heading back towards stability and predictability guaranteed by the sea wall, I pondered this dissonance. I thought of how walking and listening practices I have been committed to for many years – as an artist, researcher, and a human being – are so often about grounding one's senses, establishing a form of belonging to the surrounding environment without necessarily questioning one's eligibility to be there. Anja Kanngieser and Zoe Todd describe this awareness as a sense of *attunement*. It is a close and generous attention to the surrounding environment that acknowledges one's position as a guest, trespasser, and colonizer (Kanngieser and Todd, 2020). Many soundwalks I have taken part in aimed to facilitate a seamless state of being in the moment, in the here and now. In contrast (or in addition), through my engagements with Vancouver's intertidal zones

DOI: 10.4324/9781003193135-12

I asked how the common paths along which we, sound scholars, artists, amateurs, sound enthusiasts, and listeners, tend to think, walk, and listen, can be brought to attention and, potentially, rerouted. How can we turn soundwalking from being innocent into a more critical practice, a technique for reconsidering how we, individuals and human collectives, situate ourseleves in and account for surrounding environments? Consequently, soundwalking became to me an exercise in not so much destabilizing the human position on the planet, but a task of differently positioning ourselves within the complex interplay of its diverse scales, temporalities, and agencies.

My earlier work with sound and other media has already explored how our state of being in a moment is afforded and negotiated by numerous relations (Smolicki, 2017; 2018; 2021). Emerging (or enforced) in the past, some of those relations continuously ripple into the present and subsequently inform and diffract our futures (Barad, 2007). Soundwalking through Vancouver's coastline further reinforced my interest in finding relations between the site's diverse temporalities; between its *timescapes*, which is how Barbara Adam describes a multidimensional perception of time that works outside and beyond the reach of our senses (Adam, 1998). Among several perceptual shifts that affected the process of composing *Intertidal Room*, a piece I will discuss in this chapter at length, the aforementioned accident was a tipping point. This moment of unanticipated instability and sudden collapse made the experience of the ground under my feet strangely unsound. It turned my soundwalking into *unsoundwalking*. That feeling has remained ever since. Instead of overcoming it, I let it stay with me. I invited it into other initiatives, including my work with the soundscapes of the Swedish North, which I will also discuss in this chapter.

Concerns and ethical considerations that emerged through these pieces have in turn informed the (re-)composition of my approaches to listening to the surrounding world, as an artist, scholar, or, simply, a listening subject. By implication, these *field-responsive* pieces, as I have come to describe them, become invitations for other listeners to engage in a similar task of recomposing how we listen to the surrounding soundscapes.[3] Consequently, in this chapter I will be regularly returning to the notion of composition, also through other, closely related terms such as decomposition and recomposition. Moving beyond their dominant associations with the art of textual and musical expression, I will connect them to environmental humanities (Tsing et al., 2017), political ecology of sound and philosophy of technology (Devine, 2020). With the concept of soundwalk composition, a term that has already been used in the past (Westerkamp, 1974; Jordan, 2013), my intention is to expand it as a sonic mode of connecting with and giving justice to the complexity of a place while walking through it. With this idea, I intend to explore how walking, listening to and composing with multiple scales, agencies, and temporalities that constitute one's surrounding environment, might simultaneously become an exercise in (re-)composing one's position within it.

Moving *with* soundscapes

Many soundscape compositions are conceived to be perceived from a relatively fixed position, regardless of the number of channels they are distributed through.

The listener's body is intended to remain relatively immobile, whereas the sounds, mobilized to travel across channels, aim to immerse the listener, or wrap them, in a dynamic sonority of a given environment, a technique known as spatialization (Hagan, 2017; Stankievech, 2007). In other words, what characterizes many soundscape compositions is an attempt to simulate movement, to represent it through aesthetic techniques, editorial manipulations, and effects, to compositionally restore the intrinsically dynamic experience of environment, in great part immobilized by the process of recording it (Akiyama, 2010). The role of mobility in soundscape composition (walking, in particular) is often limited to the phase of collecting sounds and sonic impressions, typically in the initial phase. Thus, when thinking soundwalking composition, my intention is to put more critical emphasis on the significance of walking and listening, throughout the entire process of composing and engaging with soundscapes.[4]

My emphasis on careful (and caring) movement through places while composing with their soundscapes (or simply by being present therein in case mobility is impeded), does not aim to better capture these places' diverse constituents. Instead, I perceive movement as a more corresponding mode of alignment with the site's inherent vibrancies; a form of accounting for its potential (or, rather, the movement's potential) to destabilize one's deeply ingrown and involuntarily inherited certainties about it. This is why, moving and composing *with* soundscapes (as opposed to composing soundscapes) requires particular attention to one's situatedness in the environment one engages with. Subsequently, it calls for more attention to *how* one navigates through them (or is present within) and *how* one invites (or not) others to move through and pay attention to the multitude of soundscapes' layers and rhythms. Ultimately, to compose with soundscapes is to remain open to a possibility of being continuously recomposed by them.

Beyond the dominant approaches to soundscape composition, there certainly are more idiosyncratic modes that acknowledge the itinerant type of participation of both composers and listeners in working with soundscapes. In those perspectives, sounds that studio-based soundscape composers would typically sample out from natural environments, neglecting their contexts, are instead approached as deeply integral elements of a soundscape composition, one that is already out there, in the environment. Barry Truax has written extensively about the context as a crucial component of soundscape composition, and how knowledge and awareness of space the sounds come from informs the listener's experience and meaning-making processes (Truax, 1996, 2017, 2018). Hildegard Westerkamp has advocated for the practice of listening in which soundscapes compositions are to be sought within the environments one moves through (2021). There, as R. Murray Schafer suggested earlier, soundscapes 'speak for themselves, in their own voice, in their own time' (1993, p. 142). From this perspective, the practice of soundwalking that Westerkamp committed herself to in the 1970s and has practiced ever since, can be seen as a partial reversal of approaches to composing in which environmental sounds are subject to extraction and then transposition into the studio/lab/multitrack software. To go on a soundwalk is hence to abandon the studio, to question and destabilize its dominant role in the process of composing. It is to contest and demysticize the perception of the studio as a place wherein, aided by tools, the composer's genius impregnates sounds with meaning and order.

Recently, this kind of itinerant approach to composing with soundscapes has been rippling into work of others: musicians, cultural producers and sound artists (e.g. Katt Hernandez, Daniela O'Fee, Giorgio Magnanensi, Tim Shaw).[5] One notable example is the practice of EcoSono put forth by Matthew Burtner, an eco-acoustician and composer whose music and research explores embodiment, ecology, polytemporality, and noise. EcoSono uses technology 'to link human and environmental expression, in an attempt to define a collaborative and symbiotic relationship between humans and the natural world' (Burtner, 2011, p. 234). Burtner stresses that EcoSono is not only a new kind of sound practice, but an institution, and, I would add, a worldview, a type of acoustemology (Feld, 2015), 'that shifts the focus of environmental art-making towards the activism of engagement' wherein 'the process of environmental interaction overshadows the resulting media' and the artistic value 'is not in the mediation but in the experience' (2011, p. 243). Echoing Bernie Krause (2002), in putting forth EcoSono, Burtner pursues a critique of the audio recording equipment, suggesting that the sample that the recorder generates, while separating the sound from the experience, 'creates the false impression that the most important sonic aspect of the adventure is contained in the audio sample' (2011, p. 236). To oppose this, Burtner deploys recording technologies as means to augment the environment in real time, in situ. The record button is pushed to the background. What is emphasized instead is the recorder's ability to transmit and amplify sound in real time.

The form of the itinerant approach to composing with soundscapes speaks to Westerkamp's and Burtner's practices. In my approach, however, I do not aspire to set recording versus transmitting, or immediacy that the real-time transmission enables, against recollection that recorded sounds facilitate. Instead, following the premise that the past is always folded into the present moment, I engage and compose with both modalities. I accept their constraints and affordances. I do this by building mutual relations between how these two disparate modalities convey and enable the transversal experience of time and place, past and present (Smolicki, 2021). Arguably, in conventional approaches, sound artists and composers work with sounds as individual units and only after assembling them into complex configurations recognize them as compositions. But aren't sounds themselves always-already compositions that span different temporalities, agencies and scales? In other words, aren't sounds that our perception is trained to single out from larger audible contexts, always already composed, deliberately or accidentally? In the following sections I will turn to two of my soundwalk compositions to address these questions. I will discuss how walking through and listening to specific geographies I engage with helps reveal how they are already composed of, and keep composing relations between, multiple histories, agencies and temporalities. Subsequently, I will discuss how these processes of composing can in turn be articulated to broader audiences through site- and time-responsive soundwalk compositions.

Recomposing with tides

The realization that any sonic event is inherently a composition – in how it is inseparable from a larger audible context it is situated in – became very clear to me upon one of my early visits to Vancouver's intertidal zones, the place already introduced at the beginning. A seemingly inconspicuous event during that visit inaugurated a series of wanderings, composing gestures and listening situations that eventually informed the *Intertidal Room* soundwalk composition.[6] On that day, walking on an extra strip of the beach revealed by low tide, I stopped, pulled my hydrophones and dipped them in one of numerous tidal pools left behind by the withdrawing water. I used my recording gear primarily to amplify my sonic witnessing of the events presently unfolding in the pool. After a few minutes of silence, I began to hear a delicate rustle. What I was ear-witnessing, without seeing it, was a limpet rhythmically scraping off the overgrowth of algae on a partly submerged rock (Figure 11.1).

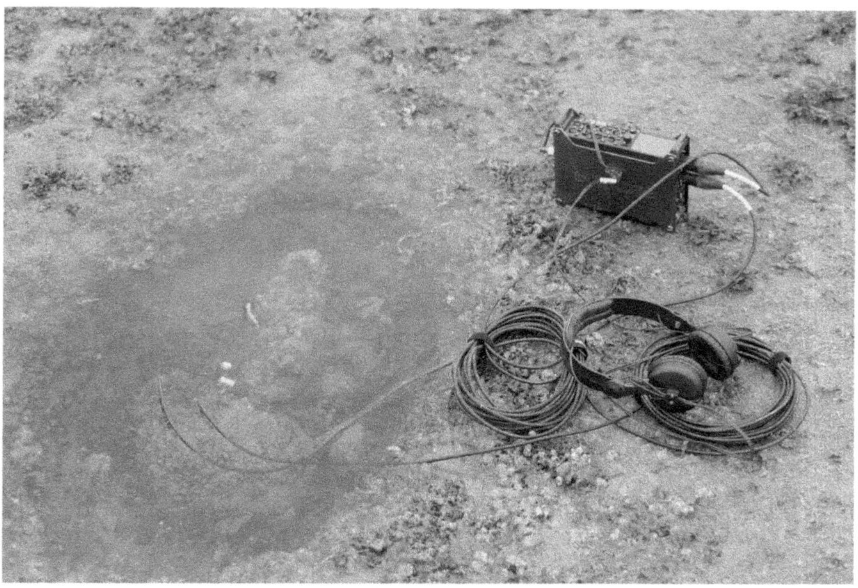

FIGURE 11.1 Listening to molluscs and other tidal organisms in one of the pools during low tide near Second Beach in Vancouver. Photo: Author, Jacek Smolicki

Limpets are small aquatic snails characterized by a conical shell and a strong, muscular foot.[7] The latter allows them to adhere to a surface of a rock and withstand the force of oceanic waves. They are often described as maintainers, gardeners, or engineers of tidal pools. This is because they systematically clean rock surfaces, in turn enabling inhabitation by other species such as barnacles and mussels. Because of this cleaning function, they are extremely important in cultivating the ecosystemic balance across intertidal zones. Additionally, through grazing overgrown algae, they maintain the temperature of oceanic waters. To scrape off the algae, limpets use a set of extremely strong teeth bristled along their tongue. What makes their teeth particularly strong are distinctive composite nanostructures consisting of high volumes of goethite nanofibers, which are considered the hardest material on our planet.[8] Besides helping them effectively scrape the algae off the rock surfaces, the limpets' teeth also provide them with mechanical stability. Metaphorically speaking, their teeth function as additional feet.[9]

More research into the life of limpets and their teeth revealed to me that geothite was named after Johan Wolfgang von Goethe, who, besides literature, was deeply committed to observation-based studies of rocks and minerals. These observations helped Goethe contribute to both sciences and arts with the concept of "primal phenomenon" (*urphänomen* in German). With it, he argued that there exists a key phenomenon to which all associated effects can be attributed (Bortoft, 1996; Goethe, 1999). Recently, Goethe's theory has been applied in marine biology, including studies of intertidal zones. Organisms that occupy them can be interpreted as organic manifestations of constraints and affordances generated by tidal variations, an instance of Goethe's primal phenomenon. Peter W. Barlow from Bristol University exemplified this by discussing how, after being taken out of their native environments and put in a lab, tidal crabs continuously follow the luni-solar rhythm that sets tides in motion (Barlow, 2012). Wondering whether their behavior derives from the learned pattern stored in their memory, or is encoded deeply in their genes, he concluded that the biological rhythms of intertidal organisms 'are not manufactured within the respective organisms' (which is to say they are not like memories) 'but seem more likely to be inherent to them, manifesting compliance with the rhythmic orbital motions of Earth and Moon around the Sun expressed in the gravimetric tide' (ibid., p. 436).

From this perspective, what I ear-witnessed in that tidal pool was not just a sonic emanation of a limpet's cyclic routine. Rather, that event, or the pool in which it took place, was a microcosm of resounding rhythms and forces whose origins reach far beyond our planetary scope and familiar temporal frameworks. That observation inspired me to keep seeking such connections between other immediate sounds and more macroscopic events happening not only in the present moment, but also in the past, across deep layers of geological time. Following this encounter, I engaged in what Anna Tsing describes as "arts of noticing," a form of recomposing our attention so it encompasses other than human rhythms and perceptions of time (Tsing et al., 2017, p. 3). The encounter with the cyclic patterns of life in Vancouver's tidal pools prompted me to shift my perception towards those, often imperceptible, rhythms although without losing touch with those persistently dominant. Upon this juxtaposition of rhythms, those anthropogenic ones became even sharper and stranger. Overall, this shift brought to the fore the complexity of relations and interdependencies between human and other than human ways of perceiving and inhabiting time and space. It helped highlight how these entanglements are inevitably expressed in soundscapes.

These relations became particularly vivid in one of my following soundwalks, this time, however, during high tide. Dipping my hydrophones in one of the pools partly reunited with the ocean, I could no longer hear that microcosm of organic pulsations generated by limpets, mussels, clams, barnacles, skeleton shrimp, and other marine organisms. What slowly began pervading my ears instead was a generic, unbearably linear hum. The noise was coming from the engines of vessels presently stationed or crossing the bay. Among some of the ships were large crude oil tankers making their way towards refineries along the Vancouver Bay. At home, using a website that tracks global movements of ships, I looked up one of them, named Atlas.[10] This crude oil tanker arrived from Rodeo in California, a stronghold of Phillips 66, the American multinational energy company involved in the trade of petroleum and the controversial construction of Dakota Access Pipeline.

Many aquatic species, most notably whales and dolphins, communicate and sense their environments via acoustic waves and hydrodynamic vibrations. Despite the lack of any hearing organs, molluscs sense vibrations through their nerves. Underwater noises, such as those generated by ships, hamper their growth and corrupt their DNA integrity (Wale, 2019). Their decline in turn leads to the increase of algae, which molluscs would normally control by preying on. Moreover, the ever-intensifying movement of vessels has been affecting the already critically endangered Southern Resident population of killer whales (orcas). In British Columbia alone, the noise from vessels takes over even up to 97 percent of the acoustic space that orcas use for social communication (Williams et al., 2014).[11] In a similar way to how I listened to the rustling sounds of tidal organisms, I began to think of this vessel's noise not in terms of a singular event but as an element of a much larger constellation. To my rewired and recomposed perception, the hum from Atlas, alongside the other vessels I listened to, became an atlas of the world's noises in which bits and pieces of very distant realms and temporalities collide, diffract against each other and dissonate. While cutting through diverse layers of marine ecosystems and affecting their cyclical rhythms at the present moment, those noises carry with them the echoes of colonial histories.

Through this kind of aural attention, that I have described elsewhere in terms of *transversal listening* (Smolicki, 2021), I learned that to truly hear, give justice to and compose with the soundscapes, one needs to look and hear back, beyond their immediacy. Therefore, throughout my following encounters with the intertidal zones, including numerous soundwalks, field recording sessions, and field-responsive sonic assemblages of different materials encountered in the intertidal zones, I sought such reverberations of the past among other seemingly uneventful yet sonically rich situations in the present moment. One example was metal detecting, a practice performed during low tide periods, predominantly by men. They would gather on the revealed strips of the beach, walking slowly in search of treasures, wedding rings, bracelets, necklaces, and coins once lost in the sea. In their actions, marked by regular beeping signals from their detectors, quite loud against the stillness of the water, I heard echoes of practices that proliferated along the British Columbia coastline in the mid-nineteenth century. At the time, hundreds of explorers sailed there from California where the gold rush was nearing the end. This rush to New Eldorado, as the newcomers called this region, posed a threat to the authorities. It mobilized them to speed up the process of establishing British Columbia. This, in turn, restrained not only the status of these new migrants but more importantly the First Nations people. Followed by extraction of other metal and minerals, while disturbing the geological composition of the land, the gold rush also rearranged relations between people, affecting social and ethnic relationships alongside issues concerning degrees of access to the land and its resources. Following this mode of transversal listening and composing with different temporalities and scales of Vancouver's intertidal zones, the canonic soundmarks of Vancouver and its coastline began to sound differently, even disturbingly if compared with the sense of cultural belonging and pride they have typically been associated with.[12] Listened from a different perspective, or a different sonic position, those "soundmarks" became sound-scars. Sound-wounds.

An array of observations of that kind, captured through field notes, field recordings and through historical research, alongside conversations with scientists, sound researchers and members of First Nations communities, eventually led to a script for the final soundwalk composition. In its narrative, the main protagonist gradually transforms from a human to an undefined element of a larger environment. What prompts this metamorphosis is the protagonist's growing openness to and acceptance of sonic registers normally inaccessible. However, this idiosyncratic skill to "echolocate," which is to sonically sense the world in a way other humans and species do (or are forced to, such as through the exposition to unbearable noises and incompatible cultural codes imposed through sound), becomes at once a poison and remedy.

FIGURE 11.2 Collective soundwalk through the intertidal zone near Second Beach in Vancouver during the premiere of *Intertidal Room* soundwalk composition. Photo: Author, Jacek Smolicki

The composition premiered in September 2020 and was available throughout the following month. The piece was live-streamed from a specially designed website only during periods of low tide. Therefore, the piece was not merely site- but also time-responsive. In order to experience it, listeners had to come to the specific location indicated on the accompanying map with their mobile phones and headphones, at a particular time of day. From there, either in groups or individually, they would embark on a slow walk along the extra strip of beach revealed during the period of so-called slack water, a moment when the tide is at its lowest. Besides the cyclically activated opportunity to individually experience the composition, I organized several group soundwalks where I gently guided participants through my experience of the intertidal zones, at times making sure that the content they listened to aligned with the specificity of the place we were at (Figure 11.2). For example, we would start by carefully walking around tidal pools, underwater sounds of which we could simultaneously hear in our headphones. When the narrative touched upon the logging industry, I would guide the group towards a pile of beached logs at the edge of the beach. Additionally, the piece was accentuated by intentional periods of silence. This was to enable participants to reflect upon what they heard as well as to let the surrounding soundscapes fade into the composition and further inform the experience of the site. The end station of those group soundwalks, about 2 km away from the starting point, was marked by a massive boulder covered by barnacles and other molluscs. After taking off our headphones, we would listen to their unmediated soundscapes and gradually shift to a conversation accompanied by kelp tea. These moments of reconnecting through conversation were often the most interesting and thought-provoking situations. It is through those conversations that the potential of soundwalking to open up to scales and temporalities, imperceptible under normal circumstances and rhythms of everyday life, was often brought to the fore. Some, mostly younger participants, found the length (about one hour), level of attention and relatively slow rhythm of this soundwalk quite challenging, especially when compared with the pace of other pedestrians walking and jogging along the nearby sea wall. Others found these aspects to be of key importance. In their view, the soundwalk provided conditions for breaking from the spatio-temporal ramifications of modern everyday life, which many of us take for granted. For one participant, the soundwalk facilitated a moment of synchrony with the "complicated temporalities" and "entangled phenomena of the surrounding environment." For others, *Intertidal Room* became what the title itself suggested: a cyclically activated room for recomposing our relationship with this specific environment in accordance to its terms.

Lake that Glimmers Like Fire

As demonstrated though *Intertidal Room*, developing a long-term relationship with a place and its timescapes is an important component of how I approach composing with soundscapes. My intention is to maintain and cultivate meaningful relationships between the site and the way it reveals itself through sound. I do this by letting the soundwalk composition stay, reverberate in, reveal the imperceptible primarily in places where I first encountered it, as in the case of *Intertidal Room*.[13] In other cases, I do it through acknowledging the soundscapes' situatedness and carefully setting the piece in relation to other spatio-temporal contexts. Through this compositional gesture of transference, I intend to deepen the story that those sounds convey, to open up other layers of those recorded soundscapes and orient listeners, at times speculatively and other times more concretely, to other scales and processes that those soundscapes might be implicated in, despite geographical and temporal distance. This was my mode of working with the Swedish North, which eventually led to the *Lake that Glimmers Like Fire* soundwalk composition.[14]

The year 2021 marked ten years of my renting a small studio at Platform Stockholm, an art collective located in an industrial area in the south of the city. To travel there, I would regularly pass by Trekanten, a small lake surrounded by a park. Since the 1960s, the lake has been known as one of the most polluted in the region. The pollution was caused by an array of industrial establishments in the nearest vicinity. As a result, Trekanten's water became heavily contaminated by large amounts of phosphorus, lead, and copper.[15] In recent decades, after pumping out some of the toxic sediments and replacing them with fresh water, the authorities have managed to improve the lake's condition.

On one of my walks around the lake, I stopped to listen to its underwater soundscapes. After dipping my hydrophones, I heard barely anything. During many following walks around the lake, I repeated my attempts, although with similar results. Only on exceptionally warm summer days was I able to hear sounds of photosynthesis, mostly from so-called "invasive plants" occupying the lake's edges. Occasionally, I was able to hear underwater insects and, sporadically, fish vocalizations, a result of systematic reintroduction of several species, including perch and pike. Against these rather delicate and scarce emanations of underwater life, Trekanten's soundscapes have been primarily pervaded by anthropogenic sounds. Resonating through the lake's bedrock and the rocks that surround it, these noises came from construction sites nearby: squeaks and clatterings of subway trains from the nearby station and a constant hum of traffic on the bridge nearby. While sitting by the shore, I wondered about this highly constructed nature of Trekanten's underwater soundscape. The problematic distinction between the natural and artificial (or natural and anthropogenic sounds that informed many early soundscape studies) did not at all hold here. The specific configuration of aquatic species in the lake (or their scarcity, rather) was a result of human decisions. Paradoxically, numerous species of invasive plants encroaching on the lake's surface, and while doing so persistently transgressing human decision-making power, seemed somewhat more "natural" in this context.

During my walks along the lake shore, followed by reading about its history and high levels of pollution, I wondered what body of water in Sweden might be considered the clearest and purest. The answer was Trollsjön, a relatively small lake located in the Swedish Arctic Circle, a region known as Sápmi, traditionally inhabited by the Sámi people.[16] In recent years, the popularity of Trollsjön (Rissajaure, in Sami language) has grown to the extent that, in 2019, after a national survey, Swedish Railway announced it as the best tourist destination in the country.[17] In summer 2021, I took a train to visit the lake. My intention was to engage in a series of listening and field recording sessions in its direct vicinity and, while doing so, set its soundscapes in relation to those of Trekanten. I stayed for about 24 hours by the lake, spending a night nearby in a secluded shelter that serves as a unit of the Swedish Polar Research Secretariat in Abisko. To listen to Rissajaure, I began by simply walking around its shore. At some point, I dipped my hydrophones into its crystal clear depth. In contrast to Trekanten, where anything that one submerges in its murky waters quickly disappears from sight, I could easily follow my hydrophones as they slowly sunk to the bottom. The lake responded with an immense silence. The only event I could hear was a delicate hiss of distant cascades carrying the water from the glacier on top of the surrounding mountains and eventually conjoining in the depths of the lake. The sound inevitably called to mind the hiss of the traffic on the bridge near Trekanten.

One part of my listening session with Rissajaure included working with a pair of self-constructed underwater speakers (Figure 11.3). I submerged them in the lake to transmit some of the soundscapes from Trekanten. In this way, I intended to symbolically aquatint Rissajaure with the noisy realm of its distant relative in Stockholm along with other recordings of underwater sonic pollution I had earlier recorded in different geographies. Using hydrophones, I recorded those noises back, as augmented by Rissajaure's waters and topography. In this symbolic act, the lake became at once a listening, mediating, and re-articulating agent. To

FIGURE 11.3 Self-constructed speakers enabling underwater transmission of sound at Rissajaure lake in northern Sweden. Photo: Author, Jacek Smolicki

symbolically end my stay with Rissajaure, and to recognize my inevitable and disruptive presence, I poured into it a small drop of water from Trekanten. I refilled the tiny bottle with a few clear drops with an intention to perform a similar gesture upon my return to Stockholm.

This encounter with the silent depth of Rissajaure brought to the fore the inevitability of my and my equipment's participation in constituting the local soundscape. In fact, from the very moment of my arrival in Abisko, after a 16-hour train ride, I pondered the intensity of material resources required to make the hike to the, allegedly purest, lake in Sweden, possible at all. Starting from the clattering of my suitcase with audio equipment, I could not stop thinking of noises that laid the ground for this and similar experiences. How much noise and disruption had to be produced elsewhere to enable the recording of the local silence? To what extent am I the agent of this disruption? In my field notes, which later informed a written narrative distributed among participants of the soundwalk, I wrote:

> Flimsy, plastic wheels carry the weight of my sensory extensions, assembled after their component bits were torn from some distant elsewheres. The bulk of my sound equipment, swirls of cables and square centimeters of black boxes, will soon be shifted onto my shoulders, back, spine, and legs, and through them, onto the body of the mountains that have never awaited me. Polymerized and polycondensated oils roll through space, passively transporting the remnants of their relatives: minerals, metals, elements, and other compounds, fragmented, processed, and forcefully assimilated into techno-cultural devices that promise to make sense of it all. Or at the very least, to give it a closer look and hearing. Taken from their invisible and inaudible milieux, they are utilized to make other habitats visible and audible.

Similar to how Trekanten's soundscape seemed like a manufactured assemblage, the soundscape of the purest lake in Sweden and its surrounding has been far from pure. The initial part of the trail to Rissajaure was under construction. Noises from excavators and road work machines continued for several kilometers into the trail. They greatly colored my sonic experience of the valley, making me rethink my approach to its soundscape. Instead of filtering out all possible instances of noise and sonic disruptions on my way to Rissajaure, I decided to acknowledge, work with, and retain them as the main material I would eventually set in dialog with Trekanten's soundscapes. Instead of seeking and mediating the alleged purity of this part of the world, I turned to its noises. In the noises of excavators, mechanically disrupting and recomposing the geological arrangement of this valley to facilitate the movement of tourists, including myself, I began to hear the echoes of disruption and decomposition of distant lands from which minerals required to manufacture my audio equipment were extracted. Similarly, in the unintended rustles, occasional glitches, and static noises of my microphones and recorders, affected additionally by humidity and rain, I began to hear soundscapes of mineral extraction. Displacement, as I reflected in my field notes, is a prerequisite to placement.

Decomposition, to composition. There is no placemaking without place-unmaking. Lastly, one's encounter with silence is someone else's exposition to noise.[18]

The title of the resulting soundwalk composition was inspired by the meaning of Rissajaure in Sami language: the lake that glimmers like fire. The piece premiered in September 2021 by Trekanten in Stockholm, almost exactly one year after the premiere of *Intertidal Room*. Unlike *Intertidal Room*, the audible content of this piece was only partly pre-recorded. The idea here was to compose on the go by mixing present soundscapes of Trekanten with those recorded at and around Rissajaure. To begin, a group of registered participants gathered in the indicated spot near Trekanten. After handing out wireless headphones, I briefly introduced the premise of the piece and technicalities. My intention was to save words to the end and let the soundscapes speak. Thus, in contrast to *Intertidal Room*, the *Lake that Glimmers Like Fire* did not interweave any spoken narrative. Instead, a written essay based on my field notes was handed to participants in sealed envelopes at the end of the walk. They were instructed to open and read (or discard) it at any convenient time after getting home. Equipped with a broadcasting kit that partly draws on Tim Shaw's soundwalking infrastructure (Shaw and Bowers, 2020) – a set of microphones and makeshift technologies, also those enabling me to transmit commonly inaudible frequency spectra, such as electromagnetic fields including those emitted by my own equipment – I invited participants to take a walk with me around the shore of Trekanten. While recording, looping, and broadcasting the surrounding sounds via a portable radio transmitter directly to participants' headphones, I was simultaneously fading in lightly processed soundscapes of the Swedish North. In result, what participants were walking through – and with – was a hybrid space constituted by two geographically distant yet closely related entities, both disrupted by human activities despite their greatly different conditions. What we were simultaneously navigating through was the uncertain space of the soundwalk composition-in-the-making, assembled in real time as we walked, and limited to the spatio-temporal conditions of this situation. At the final moment of the walk, to complete the gesture initiated by the shore of Rissajaure, I stopped to pour a few drops from its crystal waters into the foggy depth of Trekanten. The conversation that followed the premiere differed from the one after the *Intertidal Room* soundwalk. To a large extent it focused on technologies involved in the performance. As opposed to the *Intertidal Room* soundwalk, which made use of participants' own media players and mobile phones, this piece had its special technical infrastructure which effectively drew participants' attention. They wanted to learn more about the set of microphones used in the piece as well as the devices that enabled the wireless transmission and mixing of the material. Moreover, due to the lack of spoken narrative in the piece, the soundscapes that participants listened to during the walk might have sounded slightly too abstract for constituting direct points of reference and material for discussion. Only a few of the participants returned to me after reading the accompanying narrative, which made me reconsider the idea of including parts of it in the future presentations of this soundwalk composition.

Conclusion

In their critical study of pollution as an effect of colonialism, Max Liboiron writes that "our ontological relationship to land is a condition of our embodied subjectivity whether that relationship is through private property, Resource, kin, or Land" (2021, p. 77).

One's relation to the environment, and hence awareness of one's positionality within it (Robinson, 2020), are aspects that cannot be ignored when composing with soundscapes. Every act of walking and listening entails an enactment of relationships to the land. Soundwalking is at once an enactment of one's habits, beliefs, preconceptions, and attitudes towards the land as well as a form of acknowledging other kinds of relationships to and within it. Soundwalking is hence not innocent. It exerts certain relations on the environment while opening up space for rethinking them.

Soundwalking composition, as presented in this chapter, attempts to recognize those problematic relations that organize a given place (and one's presence within them). It seeks other ways of relating to it. This attempt is pursued through making room for a heightened state of aural sensitivity. However, one needs to be aware that by attempting to reconfigure one's perception and position in a given place, soundwalking – as a research method, form of activism, or aesthetic medium – might, accidentally (or not), sustain some of the problematic relations with the environment. As Liboiron suggests,

> research and activism, scientific or otherwise, never happen on a blank slate. As a result, we are always caught up in the contradictions, injustices, and structures that already exist, that we have already identified as violent and in need of change.
>
> *(2021, p. 134)*

For example, in the discussed projects, although aware of audio technologies' implication in extractivist industries and environmental injustice, I continuously rely on them. Similarly, I keep asking myself about my right to navigate through and work within places my projects are concerned with. To what extent am I invited and allowed to soundwalk through them? And yet, I still do it. What these diverse engagements with places taught me is that reconfiguration of one's orientation to the subject of concern is always partial. Some habits and attitudes that one brings along are always to some extent inevitably conserved. Nevertheless, as opposed to shielding oneself from those moments of troubling realizations, irreconcilable tensions, and dilemmas, or completely withdrawing from the field, I believe that one can approach them as productive points of departure. One can recognize them as specific (perhaps only) gateways, tight and uncomfortable and yet helpful in facilitating initial steps into the environment one intends to enter and genuinely work with.

As a conclusion, I put forth several suggestions for how to engage in an itinerant mode of composing with soundscapes I presented in this chapter as a soundwalk composition. These observations are not at all prescriptive. Similarly, they do not exhaust the case and hence should not be treated as a model reproducible and applicable across diverse sites and contexts. Rather, they can be heard as propositions for how working, walking, and living with soundscapes might be recomposed considering today's ever intensifying tide of uncertainties: environmental, social, and technological.

- Soundwalk composition starts and ends with questions concerning one's position in the multidimensional complexity of the environment one navigates through. It also invites others into this process. Compositional work with a site in each and every case requires specific and careful response, one that begins from uncertainties rather than assumed certainties about the place. Or rather, one that begins from questioning the assumed certainty of one's position within its soundscapes. In practice, this means acknowledging that one's embodied subjectivity and sensorium are constituents of only one position among a network of others through which the surrounding space is experienced, given meaning to, walked through, and materially affected. We do not merely walk through soundscapes, but soundscapes walk through us.
- Soundwalk composition cultivates relations to the place and its diverse temporalities. It limits conditions and time during which sounds are being immobilized, isolated from their fields of relations, and brought to controlled acoustic environments. This commitment is actualized in an intentional limitation of workload undertaken in a studio environment for the benefit of increased attentiveness *to* and *in* the actual environment. In soundwalking composition, the experience of environment from a position of a soundwalking body is hence the main mode of working with and articulating the complexity of spatio-temporal relations. In practice, respecting spatio-temporal aspects can be actualized through situating and sharing the work primarily (if not only) in the place of concern and in close correspondence to its distinct temporalities (such as during low tides, as in the discussed case of soundwalking through intertidal zones).
- Soundwalk composition draws on a transversal approach to the site and its soundscapes. It means that in contrast to many common ways of engaging in soundwalking – those facilitating a state of immersion in the moment – soundwalk composition attends to listening as a form of engaging with different temporalities. It transversally cuts through the time axis, (re-)composing links between present and past events. In other words, in soundwalk compositions, listening becomes a relational practice in which present events, as expressed in soundscapes, are tightly connected to past ones, even those in deep geological time. By implication, what we hear today is in part premonition of what is to come. To facilitate this movement across the past, present and future, soundwalking composition draws on diverse forms of mediating sound that in turn bring attention to different temporal aspects: real-time transmission, unmediated listening to the immediate surroundings, field recordings and archival material.

- Soundwalk composition makes room for critical and ethical reflection on material and technical infrastructure it is inherently dependent on. From the media-material perspective, any soundscape composition made in an organized acoustic environment (and intended for such) always carries its imprint. Any composed soundscape represents its technical environment – its limitations and affordances – as much as the places from which its sonic ingredients were extracted. Like composed soundscapes, soundwalk compositions are to some extent *schizophonic* constructs (Schafer, 1977, p. 90), even before being listened to by the audience. Their very substance does not relate to a single environment only, but multiple ones at once. A single soundscape composition is hence a conglomerate comprising multiple other technical, cultural, and environmental composites: field recordings, filters afforded by specific software environments, chemicals and minerals needed for the production of audio equipment, etc. (Smolicki, 2022). Soundwalk composition, as any soundscape composition, is hence an unintended carrier of other relations. Within itself it carries elements of integrity and rupture, containment and leakage, compositional consistency and spatio-temporal inconsistency. Site-specificity and deterritorialization. However, in contrast to soundscape composition in which those relations are often kept invisible, soundwalk composition attempts to acknowledge and critically work with them. In practice, this might include raising awareness about resources that one's technologies are dependent on, acknowledging human and other than human realms that the production of those technologies disrupts and, consequently, a proactive turn to other technologies, second-hand options and maintenance practices. On a higher, ontological level, the soundwalk composer might ask, after Anna L. Tsing, how can the recording tools of modernity be tactically "repurposed against the terror of progress" in which they were originally implicated? How, in turn, can they be used to make audible the other worlds that modernity has ignored and damaged (Tsing, 2017, p. 7)?

Notes

1 This chapter stems from my 3-year international postdoctoral research entitled 'Soundwalking. The Art of Walking and Listening through Time, Space, and Technologies', funded by the Swedish Research Council (Vetenskapsrådet), between 2020 and 2023.

2 Stanley Park was established by settler colonizers in 1887. In contrast to the official rhetoric of the founders of the park presenting the land as previously unoccupied, it had for many centuries before been home to multiple indigenous settlements (Kheraj, 2013; Barman, 2007). The most prominent of those was X̱wáýx̱way (also known as Whoi Whoi).

3 I intentionally use the term "field-responsivity" as opposed to the more commonly used "site-specificity." Replacing site with the field points out that specificity of a place is not contingent on its physical location alone but also the place's entanglement in a mesh of spatio-temporal relations (with other places and agencies) that require attentive, responsive, and responsible engagement over a period of time.

4 I am aware that this emphasis might imply the ability of a body to walk and thus risks excluding differently-abled bodies. I am in agreement with scholars who have rightly pointed out the limitations of soundwalking in this respect and flagged the deficit of

perspectives acknowledging this as a bigger problem within arts and culture (Rose, 2020). I am aware that the immobilization of environmental sounds I present as dominant in soundscape composition might be seen as problematic, especially from the perspective of a body that is able to walk, such as mine. For the differently-mobile body, the mobilization of sounds through editing, mixing, and effects might in fact be the only way to sense the vibrancy of a specific environment.

5 For more examples, see Walking Festival of Sound (www.wfos.net), a transdiciplinary event I co-founded with Tim Shaw in 2019.

6 *Intertidal Room* was developed in 2020 as part of my research visit to the Sonic Research Studio/School of Communication, at Simon Fraser University at the invitation of prof. Barry Truax.

7 Peter Godfrey-Smith draws a genealogical connection between limpets and some of the earliest mollusks on our planet appearing in the Cambrian era. During this period, starting about 541 million years ago and lasting for the following 55 million years, mollusks developed shells and, some of them, a muscular foot allowing them to crawl along oceanic bottoms. Gradually, the dominant function of the muscular foot was rendered unnecessary, and the foot began its slow transition into tentacles. This transformation enabled mollusks to better control their movements and gradually turned them into cephalopods, at the time the most fearsome predators in the sea (2016, p. 46).

8 Geothite is a mineral built of iron oxide-hydroxide.

9 The strength oflimpets' teeth was found to range from between 3.0 to 6.5 GPa (Gigapascals), which is the highest recorded for a biological material on our planet, outperforming spider silk, long considered to be the strongest natural material (Barber et al., 2015).

10 https://www.vesseltracker.com, accessed July 15, 2022.

11 See also: Macarenko, Gloria, 2019. Killers: J pod on the brink, CBC, https://www.cbc.ca/listen/cbc-podcasts/375-radio-killers-j-pod-on-the-bring, accessed July 15, 2022.

12 Here I mean in particular the Naval cannon at Brockton Point firing every day at 7:00 pm, Heritage Horns at noon, fog horns from cargo ships, the Gastown steamclock, and the sounds of the Canadian Pacific Railway.

13 Upon special request, I would share the file with listeners located in distant places. However, they would also receive special instructions asking them to perform the walk in the vicinity of water (for example a river, lake, or pond) and, prefereably, around the time of low tide.

14 The project was developed during my research visit facilitated by the Abisko Scientific Research Center in northern Sweden in August 2021.

15 Vattenprogram för Stockholm 2000, Trekanten. Faktaunderlag. An information brochure released by Stockholm Vatten och Avfall (Stockholm Water and Wastewater). https://web.archive.org/web/20070927014333/http://www.stockholmvatten.se/Stockholmvatten/commondata/infomaterial/Sjo/14Trekant.pdf, accessed July 20, 2022.

16 While I could not verify this to be an outcome of any rigorous scientific research, the status of Rissajaure being the purest lake has been quite predominantly recognized, agreed upon and shared across numerous tourist services, local authorities, and in public opinion as expressed in online materials, articles, and blogposts. https://res.se/magiska-trollsjo-korad-till-sveriges-basta-resemal, accessed June 12, 2022.

17 https://res.se/magiska-trollsjo-korad-till-sveriges-basta-resemal, accessed June 12, 2022.

18 In his recent book, Mark Peter Wright shared similar observations arguing that "the practice of field recording is bound into the production of the field as a site of displacement" and that the practice is "built upon asymmetrical relations of power." (2022, p. 20).

References

Adam, B. (1998) *Timescapes of Modernity: The Environment and Invisible Hazards*, New York, NY: Routledge.

Akiyama, M. (2010) Transparent Listening: Soundscape Composition's Objects of Study. *RACAR: revue d'art canadienne / Canadian Art Review*, 35(1), Landscape, Cultural Spaces, Ecology / Paysages, espaces culturels, écologie, 54–62.

Barad, K. (2007) *Meeting the Universe Halfway: Quantum Physics and the Entanglement of Matter and Meaning*. Durham and London: Duke University Press.

Barber, A. H., Lu, D., and Pugno, N.M. (2015) Extreme Strength Observed in Limpet Teeth. *Journal of the Royal Society Interface*, 12, 20141326.

Barlow, P. (2012) The Primal Integrated Realm and the Derived Interactive Realm in Relation to Biosemiosis, and Their Link with the Ideas of J.W. von Goethe. *Communicative and Integral Biology*, 5(5), 434–439.

Bortoft, H. (1996) *The Wholeness of Nature—Goethe's Way of Science*. Edinburgh: Floris Books.

Burtner, M. (2011) EcoSono: Adventures in Interactive Ecoacoustics in the World. *Organised Sound*, 16(3), 234–244.

Devine, K. (2020) *Decomposed. The Political Ecology of Music*, Cambridge, MA: MIT Press.

Feld, S. (2015) Acoustemology. In: D. Novak and M. Sakakeeny, eds. *Keywords in Sound*. Durham, NC: Duke University Press, pp. 12–21.

Godfrey-Smith, P. (2016) *Other Minds: The Octopus, the Sea, and the Deep Origins of Consciousness*. New York, NY: Farrar, Straus and Giroux.

Goethe, J.W. (1999) *Schriften zur naturwissenschaft*. Germany: Reclam.

Hagan, K. (2017) Textural Composition: Aesthetics, Techniques, and Spatialization for High-Density Loudspeaker Arrays. *Computer Music Journal*, 41(1), 34–45.

Jordan, R. (2013) Vancouver Soundscape Chronicles, Bell Tower of False Creek, blog post. Available at: http://www.randolphjordan.com/schizophone/vancouver-soundscape-chronicles-bell-tower-of-false-creek/, accessed August 9, 2021.

Kanngieser, A., Todd, Z. (2020) From Environmental Case Study to Environmental Kin Study, *History and Theory*, 59(3), Wesleyan University, pp. 385–393.

Kheraj, S. (2013) *Inventing Stanley Park. An Environmental History*. Vancouver, BC: UBC Press.

Krause, B. (2002) *Wild Soundscapes: Discovering the Voice of the Natural World*. Birmingham, AL: Wilderness Press.

Liboiron, M. (2021) *Pollution Is Colonialism*. Durham, NC: Duke University Press.

Robinson, D. (2020) *Hungry Listening. Resonant Theory for Indigenous Sound Studies*. Minneapolis, MN: Minnesota University Press.

Rose, M. (2020) Access Denied: Disabled People and Walking Art. In: H. Billinghurst, C. Hind, and P. Smith, eds., *Walking Bodies. Papers, Provocations, Actions*. Axminster, UK: Triarchy Press, pp. 218–228.

Schafer, R.M. (1977) *Soundscape, Our Sonic Environment and the Tuning of the World*. Rochester, VT: Destiny Books.

Schafer, R.M. (1993) *Voices of Tyranny, Temples of Silence*. Douro, ON: Arcana Editions.

Shaw, T. and Bowers, J. (2020) Ambulation: Exploring Listening Technologies for an Extended Sound Walking Practice. *Proceedings of NIME* (New Interfaces for Musical Expression). Available at: https://www.nime.org/proceedings/2020/nime2020_paper4.pdf, accessed June 21, 2022.

Smolicki J. (2017) *Para-Archives. Rethinking Personal Archiving Practices in the Times of Capture Culture*. Doctoral dissertation, Malmö University.

Smolicki J. (2018) You Press the Button, We Do the Rest. Personal Archiving in Capture Culture. In: A. Romele and E. Terrone, eds., *Towards a Philosophy of the Digital*. New York: Palgrave Macmillan, pp. 77–100.

Smolicki, J. (2021) Minuting. Rethinking the Ordinary through the Ritual of Transversal Listening. *VIS – Nordic Journal for Artistic Research*, 5. Available at: https://www.visjournal.nu/minuting-rethinking-the-ordinary-through-the-ritual-of-transversal-listening, accessed July 15, 2022.

Smolicki, J. (2022) Acoustethics: Careful Approaches to Recorded Sounds and Their Second Life. In: S. Wieczorek and R. Tanczuk, eds., *Sensitive Sound Recordings*. Prace Kulturoznawcze: Wrocław University Press, vol. 26 I. 1, pp. 11–33.

Stankievech, Ch. (2007) From Stethoscopes to Headphones: An Acoustic Spatialization of Subjectivity. *Leonardo Music Journal*, 17, 55–59.

Truax, B. (1996) Soundscape, Acoustic Communication and Environmental Sound Composition. *Contemporary Music Review*, 15(1), 49–65.

Truax, B. (2017) Editorial. *Organized Sound*, 22(1), 1–3.

Truax, B. (2018) From Epistemology to Creativity. Personal View. *Journal of Sonic Studies*, 1(4). Available at: https://www.researchcatalogue.net/view/266108/266109, accessed August 13, 2022.

Tsing, A.L., Swanson, H.A., Gan, E., and Bubandt, N., eds. (2017) *Arts of Living on a Damaged Planet. Ghosts and Monsters of the Anthropocene*. Minnesota, MN: University of Minnesota Press.

Wale, M. (2019) From DNA to Ecological Performance: Effects of Anthropogenic Noise on a Reef-building Mussel. *Science of The Total Environment*, 689, 126–132.

Westerkamp, H. (1974) Soundwalking. *Sound Heritage*, 3(4).

Westerkamp, H. (2021) The Microphone Ear: Field Recording the Soundscape. *Review Filigrane. Musique, esthétique, sciences, société*, 26.

Williams, R., Clark, C.W., Ponirakis, D., and Ashe, E. (2014) Acoustic Quality of Critical Habitats for Three Threatened Whale Populations. *Animal Conservation*, 17, 174–185.

Wright, M.P. (2022) *Listening After Nature. Field Recording, Ecology, Critical Practice*. London: Bloomsbury.

INDEX

www.ingramcontent.com/pod-product-compliance
Ingram Content Group UK Ltd.
Pitfield, Milton Keynes, MK11 3LW, UK
UKHW020006200926
475702UK00015B/2506
* 9 7 8 1 0 3 2 0 4 4 2 2 4 *